Brooklyn Historic Railway Association

Bob Diamond
Founder, President

Greg Castillo,
Vice President, Implementation & Quality Control
Assurance

Brian Kassel
Vice President, Design & Planning

Dylan Cepeda,
BHRA Publication Editor, Primary Researcher,
Graphics Design

Brooklyn Historic Railway Assn.
599 East 7th Street   Ste 5A
Brooklyn, NY 11218
rdiamond@brooklynrail.net

www.brooklynrail.net

Is This Behind the Wall?

599 E. 7TH ST., SUITE 5A BROOKLYN, N.Y. 11218   WWW.BROOKLYNRAIL.NET

BROOKLYN HISTORIC RAILWAY ASSOCIATION

EXECUTIVE SUMMARY

The old tunnel, that used to lie there under ground, a passage of Acheron-like solemnity and darkness, now all closed and filled up, and soon to be utterly forgotten, with all its reminiscences; of which, however, there will, for a few years yet be many dear ones, to not a few Brooklynites, New Yorkers, and promiscuous crowds besides.
- Walt Whitman on the Atlantic Avenue Tunnel, 1862

Objective

A very exciting proposal is now being put forth by the Brooklyn Historic Railway Association (BHRA), a non-profit education corporation. The idea is to reopen the historic Atlantic Avenue Tunnel, the oldest subway tunnel in the world, built in 1844, improving public access and restoring the tunnel as a living museum and historic attraction. The museum will be different from surrounding institutions like the transit museum because visitors will be immersed in historical reproduction as well as a learning environment. The tunnel will be akin to colonial Williamsburg in Virginia.

The BHRA seeks to provide a unique cultural experience by preserving the tunnels history as well as fostering the development of community and social capital. The tunnel will benefit the community by providing a historical education of the neighborhood as well as transportation and technology. The tunnel will attract tourists into Downtown Brooklyn which is already undergoing a renaissance due to the increased development in the area. The tunnel which is only accessible by foot, will encourage people to walk along Atlantic Avenue and patronize local businesses. The presence of the Museum will benefit the area economically. The tunnel museum will also give back to the community by making itself available for a number events primarily aimed at local residents. These special events will promote the arts and sciences while at the same time reflect the history of the tunnel.

Project Outline

The Project has 3 major phases:
- Improve visitor access to the tunnel by installing a proper entrance, new lighting, and modern safety equipment. This will allow more people to access the tunnel and provide an effortless decent into the depths beneath the street.
- Discovery and protection of the locomotive buried in the tunnel. The locomotive has been detected by scientific equipment but has not been seen by a person in over a century. The locomotive needs to be found and preserved in order to prevent this one of a kind find from rotting away.
- A replica locomotive should be constructed. A replica will help visitors conceptualize the history of the tunnel and transportation at the time and will provide an interactive exhibit. It will also provide a learning exhibit for volunteers wishing to participate its construction.

THE EXPERIENCE

The tunnel museum will be a unique attraction in New York City and will draw residents and tourists into Brooklyn's downtown area. No where else in the city is there an abandoned tunnel that is easily accessible to city tourists. Few other cities offer a similar attraction. Sacramento and Seattle both on the west coast The tunnel will be a draw for post modern consumers.

The tunnel museum will be much more than the average museum experience. The tunnel will be designed to look as if it is still in a state of archaeological discovery. Visitors will feel as if they are entering the tunnel for the first time and embarking upon a ground breaking discovery. In this vein the tunnel will seek to draw upon the excitement of the growing urban exploration movement (urbex). Urban explorationists generally explore paces within the built environment that are off limits to the general public. As an attraction the tunnel will provide all the experiences of urban exploration without the risk. This increases the appeal of the tunnel to a broader audience. Post modern consumers will enjoy the realism of the simulated risk of archaeological discovery while knowing that they are completely safe from harm. This risk-less risk or sanitized razzmatazz is a feature of a number of successful urban attractions across the country. In addition, we hope to attract: the general public thru the "human interest" story of how Robert Diamond found the tunnel against all odds a modern David and Goliath tale, Some train buffs who wish to see the locomotive Students, researchers, and conspiracy theory/paranormal investigators (huge potential market) like these, for example https://www.youtube.com/watch?v=mXH6rj-9sZU.

Visitors to the tunnel, from the moment of entry, are engaged in a personal, emotional, act of archaeological discovery experiencing what Bob Diamond felt at the moment of discovering the tunnel. Visitors will descend into the tunnel through a new entrance and at the base of the stairs will be a dramatic stone entranceway. This portal will mark the beginning of the simulated experience. The walkway following the portal will be covered with flashing LED panels and blanketed with a layer of fog from fog machines. Visitors will be made to feel as if they are entering another dimension. The LED and fog will give way to simulated torch lighting and a mine-like walkway as visitors will realize they are now in an archeological dig site similar to those featured in Indiana Jones films. Visitors will exit this small walkway through a hole that opens up into the expanses of the tunnel. In the tunnel will be decorative piles of rubble and strategically placed artifacts that have been found within the tunnel. The first stop

on the tour will be this simulated dig site. There will also be open spaces free of decoration for special events and to show off the architectural significance of the tunnel design.

After passing through this section visitors will be transported with the aid of multimedia devices including, led lighting, video projections, fog machines, and sound. This will be the first "time jump". Visitors will then find themselves in 1861 Brooklyn. Projections and lighting will make it appear as if visitors are on the surface. Visitors will be led by the tour guide to strategic points to meet tunnel characters portrayed by actors. These characters will include robber baron Augustus Litchfield, John Wilkes Booth, the Smoky Hollow Gand, and writer Walt Whitman. Visitors will learn about the importance of these characters and how they became involved with the tunnel through an interactive exchange. Visitors will get to ask questions and talk with the characters and actors will respond in the voice of the characters. Actors will also interact with each other to portray events that took place in the tunnel or reenact local tales of what was believed to have happened in the tunnel.

A second time jump will transport visitors into the early 20th century. There they will meet a new set of characters including H.P Lovecraft, WW1 German Spies, Prohibition era bootleggers, and ghosts. A similar set of interactions will occur in this time jump. However, this jump will have more occult themes such a vampires and ghosts. It will be a popular stop during the halloween season.

Eventually visitors will jump back to the present. They will reach a wall and believe that the tour is over however it will not be. By pressing a button, the guide will open a "secret" door disguised by hologram which will open into another mine like tunnel. After passing through visitors will reach the engine room which contains the lost locomotive. Dramatic spotlights will reveal the locomotive in the dimly lit room. Visitors will be able to act as if they have found the archeological find of the century. Finally, visitors will have photo opportunities with the engine and actors before leaving through a secret entrance below a storefront.

599 E. 7TH ST., SUITE 5A BROOKLYN, N.Y. 11218   WWW.BROOKLYNRAIL.NET

Inspired just a tad by an original Star Trek episode on time travel paradox, and partly by the unique imagery of French renaissance artist Hubert Robert, the Atlantic Avenue Tunnel Tour combines the numerous "truth is stranger than fiction" stories surrounding the Atlantic Avenue Tunnel. Visitors will be transported to other times and realms of "imagineered historical accuracy".

The Interior of the Temple of Diana at Rimes, Hubert Robert, 1783

Plundering the Royal Vaults at St. Denis in October 1793

The crew of the Enterprise landing party encounters the Time Portal known as the "Guardian of Forever". From Star Trek: The Original Series episode, "The City on the Edge of Forever" *(SE01, EP28)*

TUNNEL VISION

Tunnel visitor's first experience is entry to the tunnel through a mysterious, long abandoned stairway, said to have last been used by actual New York Harbor "River Pirates" known as the "Smoky Hollow Gang" during the 19th century.

The background is set off by a variation of the existing stone tunnel bulkhead wall under Hicks Street, based upon the Hubert Robert painting "The Temple of Diana at Nimes". A projected 3- dimensional holographic image would initially conceal access to the main body of the tunnel. The wall then transforms into a "Time Portal", and the "walking experience" through the tunnel begins. Heavy use of 3- dimensional holographic imaging and sound reinforcement throughout the tour is anticipated. A special "Tunnel Vision" video game is in the planning stage, and models of scenes will be available at the gift shop.

The openings broken through the stone block wall, and the ghostly human form seen emerging, sets the tone of the entire "tunnel experience" - breaking through the "barrier of time and space", and passing into other times and worlds of possibility.

This diorama model depicts how we hope the locomotive in the AA Tunnel is situated on the other side of the Hicks Street bulkhead. Ideally, there's an open section of tunnel behind the wall, but it may also be piled up with varying degrees of backfill too, that would require digging out.

BRIEF HISTORY OF THE TUNNEL

An ordinance of the Brooklyn Common Council dated March 29, 1844, granted authority to the Long Island Railroad to construct the Atlantic Avenue Tunnel. The railroad planned to use the Tunnel as a major artery in their rail service between New York and Boston. This rail line was part of a much larger system of railroads that extended from Boston to Charleston, S.C. The Tunnel was a major breakthrough in transportation technology and city planning. It carried trains under Atlantic Avenue, thereby preserving the then fashionable shopping street and its inherent pedestrian and vehicular traffic. It was the prototype of "cut and cover" subway construction, the method still used today, in which long trenches are dug in the street and then covered to form the tunnel corridors. The development of this process had an historic impact on urban planning and development; it enabled planners to integrate railroads into complex urban landscapes and led directly to the creation of metropolitan subway systems.

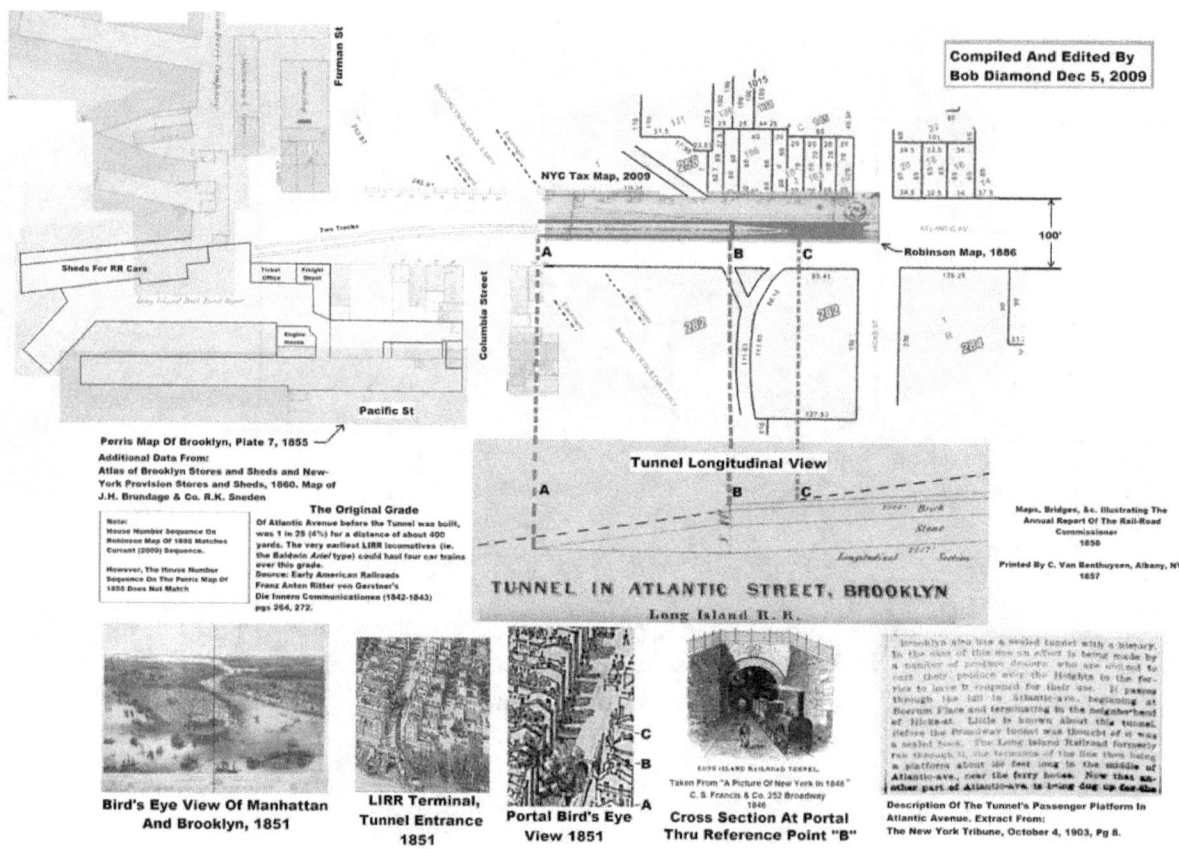

After the Tunnel was completed in 1844, Brooklyn became a major transportation and commercial center to rival New York, and grew to be the third largest city in the country (a distinction it held until 1898 when it became a borough of greater New York). In 1848, competition from New York in the form of the New Haven Railroad caused the LIRR to lose its monopoly on rail service to Boston, and led to substantial financial losses and the abandonment of its interstate service.

Only a few years later a prominent developer, Mr. Electus Litchfield, schemed to close the Tunnel and remove the LIRR from Brooklyn in order to create an Atlantic Avenue Boulevard and Promenade, a grandiose project inspired by the Champs-Élysées in Paris. With the help of corrupt politicians, Litchfield pushed the illegal legislation which permitted him to tax Atlantic Avenue merchants and property owners for the removal of the Tunnel and the LIRR, which he had branded as a "public nuisance." As a result, steam locomotives were banned in Brooklyn in

ATLANTIC AVENUE, DRIVE AND PROMENADE.

1859 and the Tunnel was finally closed and sealed in 1861. In only a few short years the Tunnel had gone from a technologically advanced project which would benefit all of Brooklyn, to a scapegoat for the corrupt plans of a robber baron. Litchfield then used the ill-gotten money to initiate his real estate project in what would become Park Slope, and build a new rail line from Jamaica to Hunters Point, the line the LIRR presently uses. However, no Boulevard was built due to the ensuing lawsuits brought by the merchants and property owners against Litchfield. The elimination of rail service left downtown Brooklyn in economic chaos, causing it to be transformed from an economic rival of New York to one of its most prized and diversified residential areas.

For over one hundred years, the Atlantic Avenue Tunnel remained sealed and largely forgotten, the subject of fantastic myths and legends which seemed to crop up with each generation-- many of which turned out to have some truth. Despite the recurrent rumors, numerous attempts to locate an entrance had failed. Finally, in early 1980, Robert Diamond first heard of the legendary tunnel on a radio broadcast about The Cosgrove Report, which claimed there was an old steam locomotive buried in a forgotten tunnel in downtown Brooklyn. The book also mentioned a legend that the missing pages of John Wilkes Booth's diary had been hidden there. Intrigued, Diamond spent seven months researching the tunnel's history, eventually locating an

unmarked manhole in the middle of Atlantic Avenue and Court Street he was sure would lead to the long-abandoned tunnel beneath. Yet when the manhole was opened, there was nothing to be seen but a three-foot drop. The dirt fill came up to about two feet from the underside of the pavement. Diamond knew at that moment he was standing on a backfilled portion of the tunnel. Looking around with a flashlight, he noticed what appeared to be a wall some seventy feet to the west. He was separated from this wall by a crawlspace less than two feet high. For the next year he searched the area, pleaded with skeptical, sometimes indifferent officials, researched, probed and slowly raised the curiosity of enough influential people to continue the exploration. In the summer of 1981, Mr. Diamond was able to crawl the seventy foot distance to the wall where he noticed the outline of a blocked-up opening in the concrete wall. The access was sealed with brick and Belgian paving blocks. After several hours of hard work with pick and shovel, Diamond and several men from Brooklyn Union Gas Company (now National Grid), who had agreed to help him on his underground mission, broke through the opening and finally saw the full expanse of the Tunnel before them, exactly as it was when sealed up 120 years earlier.
In 1982, Mr. Diamond founded a not-for-profit corporation, the Brooklyn Historic Railway Association (BHRA), to preserve and restore the tunnel, and establish a museum and scenic

railway. For the past twenty-nine years, BHRA, in conjunction with elected officials, city agencies, community groups and local businesses, has been working to develop the Tunnel as a valuable public asset. BHRA received all the necessary approvals for a franchise from the City of New York to occupy and operate the tunnel as a museum devoted to the study of early railway transportation. BHRA has also been fostering public awareness and support for this forgotten municipal treasure, hosting public tours which have been enjoyed by thousands of city residents and tourists alike. During this time Mr. Diamond has further explored the tunnel's history and its significance to New York. Because it is the earliest known example of the cut-and-cover technique of railroad tunnel building in the world, and because it was part of New York's earliest train service, the tunnel has been listed on the National Register of Historic Places since 1989. The tunnel is also recognized by the Guinness Book of World Records as the "World's Oldest Subway Tunnel", starting in the 2011 edition.

TOURISM

Since 1982, BHRA has offered both public and private tours of the tunnel which have been enjoyed by thousands of visitors. Many private and public schools have sent hundreds of students on class trips. Most recently, at the behest of the city Department of Transportation, regular public tours were reinstated in 2007 and given about twice per month through the end of 2010. During this time public interest in the tunnel and its history increased dramatically, and hundreds of people were safely led through the tunnel on guided tours given by Mr. Diamond. The Tunnel received over 12,000 visitors in 2010. Both New Yorkers and tourists from all fifty states and many foreign countries lined up for the adventure of seeing the legendary underground expanse for the first time. The response was overwhelmingly positive. Visitors reported feeling a strong sense of mystery and intrigue, as well as the sensation of traveling back in time to the 19th century. Teachers commented afterwards that students were highly motivated by the visit. Numerous newspaper and magazine articles have been written about the Tunnel, including feature stories in The New York Times, Daily News, National Geographic, Science Digest, and The New Yorker. The project has also been covered by local television and radio as well as national exposure on CNN and the Canadian Broadcasting Corporation. In 2009, the tunnel was featured on the History Channel TV show Cities of the Underworld. In addition, National Geographic has begun work on a documentary focused on the historic locomotive buried at the western end of the tunnel.

The Atlantic Avenue Tunnel Museum will be designed to appeal to the general public as well as to academics, train buffs, and conspiracy theorists. With its dramatic subterranean location and exhibits which will include historic train cars and railroad artifacts, the museum should prove of particular interest to children.

The museum will have both local and international appeal. The Brooklyn Historic Railway Association estimates it will draw at minimum 10,000 visitors per year during our proposed from the tri-state area, as well as tourists sightseeing in New York City. The museum, located beneath a busy Brooklyn thoroughfare, will also draw visitors from its immediate neighborhood, a melting pot of African Americans, Hispanics, Middle Eastern émigrés and families of Italian American descent.

Since future development in downtown Brooklyn will rely on the intrinsic assets of the area, it is the old Atlantic Avenue Tunnel which highlights the primary asset of the community—easy access and unparalleled transportation facilities. The museum, in the heart of downtown

Brooklyn, is just a short walk from federal courts, office buildings, city government offices, and the historic homes of Brooklyn Heights, the first designated landmark district in the United States. The Atlantic Avenue Tunnel can thus serve as an historic symbol for today's public and private sector leaders as they reemphasize downtown Brooklyn as a business and transportation center now, as it was 160 years ago.

This project will enhance the quality of life in an area now experiencing a major renaissance, as well as ensuring the redevelopment of downtown Brooklyn from both an economic and social standpoint. It would have a synergistic impact on several other projects currently underway downtown.

As well as providing a new cultural resource and tourist attraction for the state and city, this project will stimulate business in the many restaurants, specialty food shops, antique stores, art galleries, and other retailers in the area. In addition, the project will generate a variety of jobs in its implementation, and serve as a centerpiece for the much publicized redevelopment of downtown Brooklyn.

MUSEUM OPERATION

Once accessible to the public, the Tunnel would have immediate public benefits. Current uses would include:
1. Guided walking tours to groups of up to 50 people at a time. These tours would take place on Saturdays & Sundays from 11:00 AM to 5:00 PM. Special weekday events may be planned.
2. Cultural gatherings.
3. Site location for media productions.

Possible additional future uses as per NYC Board of Estimate resolution adopted on October 9, 1986:
1. Historical exhibits.
2. Streetcar/railway museum and/or railway vehicle storage "barn."
3. Partial use as part of a future streetcar line.

On any typical 2010 Sunday afternoon tunnel tour date, regardless of season or weather conditions, BHRA received on average, about $5,000 in free will contributions ($4,000 low, and over $6,000 high).

We base our future Revenue Projection upon past performance over the last three years, and the assumption that the suggested contribution for tunnel tours will be raised to $20 per person (a 30% increase), and that the planned improvements would allow the operation of tunnel tours/museum to be expanded to 7 days a week, with the circa-1830's locomotive discovered in the tunnel made part of the exhibit. Based upon the foregoing, we project gross revenue would be in the neighborhood of:
$6,500 per day x 360 days = $2,340,000 per year gross project revenue

PROPOSED CULTURAL PROGRAMS

The following proposals are only a small selection of special events and programs that can be held in the tunnel and future programs are not solely limited to these events. Historic tunnel tours will be offered on a regular basis.

Robert Diamond Shows His Entry To NEF President Alan L. Smith

Science Fair
The Atlantic Avenue Tunnel would have never been rediscovered if Robert Diamond had not win the science fair. Hosting a new annual science fair will help inspire children to pursue science careers.

Craft Beer Festival
Craft Beer has become a popular consumable nationwide and New York City has several brewers. The Tunnel can be used to host a beer tasting competition and even inspire a new flavor of beer reflecting Brooklyn's trolley dodging past. The tunnel was the site of historic alcohol boot legging during its operation and artifacts from these activities are extant.

Movie Screenings
Classic films, documentaries, and independent films can be shown to a limited audience in the tunnel. The tunnel can also double as a shooting location for New York's burgeoning film industry.

Live Theatre
Temporary stages can be constructed in the tunnel to accommodate performances of simple plays.

Live Music
Local performance artists can play their own compositions to a small number of people as well as original ragtime music relating to streetcars and railroads.

Transit Technology & Science
Visitors will get to examine the locomotive and learn about the history of local transit and the technology that powered it. Visitors who want to get involved will work hands on with restoration and replica construction in the museum.

THE SITE

The Atlantic Avenue Tunnel

After being sealed for over a century, the Tunnel is a perfectly preserved, truly magnificent structure. It is a half-mile long, twenty-one feet wide and seventeen feet high. Its walls are six-foot thick granite blocks and the roof is a three-foot thick brick arch. Several prominent civil engineers have been actively engaged in determining the tunnel's structural soundness and architectural and engineering significance, and have concluded that it is structurally perfect. In fact, they have compared it to the pyramids of Egypt. An evaluation performed by LMW Engineering Group, LLC, in March 2009, found the tunnel "impressively devoid of any sign of deterioration." Their report further concludes that:

- The structural integrity of the tunnel is sound and has not been compromised by aging.
- The tunnel can be considered safe under its current use for visitors and tourist attraction.
- There is no evidence that any form of maintenance or repair work is necessary at this stage.
- The tunnel can be safely, with relatively minimum rehabilitation effort, mostly esthetic, be utilized as a museum or similar facility.

In summary, the tunnel, as inspected by us, is a safe and sound structure.
Studies conducted by prominent consultants as well as by the City departments of Sewers, Water Supply, Transportation, Fire and Electrical Control, and a study by the National Historic Register have led to the following appraisal: The tunnel is a marvel of early engineering techniques, historically one of the most important architectural structures of the 19th century.

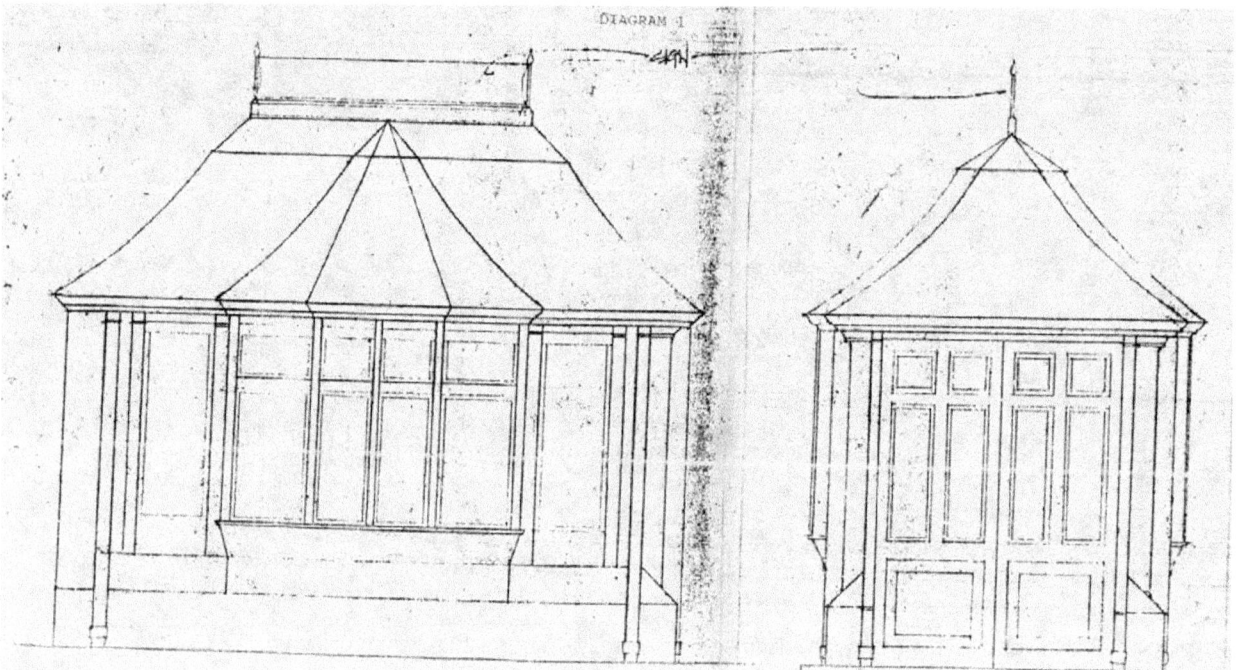

Above: Period Appropriate Entrance Kiosk

IMPROVEMENTS

The first phase of access improvements will involve the construction of a side walk entrance at Court Street and Atlantic Avenue. This entrance will be placed on the corner with the least potential conflicts with existing utilities. Existing utility maps reveal that the south east corner at the intersection of Court Street and Atlantic Avenue is the most suitable area for a new entrance. This new entrance will consists of a period appropriate entrance kiosk as well as well as modern steel stairways which will lead into the tunnel.

A second entrance will be constructed on the corner of Hicks Street and Atlantic Avenue. Utility maps reveal that the south west corner of the intersection is a suitable location for an elevator. The second entrance will utilize a modern machine roomless elevator and will be constructed on the corner with the least underground utility conflicts. This second entrance will be similar to elevators recently installed in MTA subway stations. We expect that the addition of the entrance with additional enhancements will make the tunnel ADA compliant and will allow a multitude of people to access the tunnel. We expect that the ball park figure for this elevator with ADA improvements will be around $7 million. This is the price that the MTA recently paid for the Dyckman Street Subway Station elevator.

Finally, a third entrance could also be constructed. This would be part of a street level museum which will occupy a storefront along Atlantic Avenue. Suitable locations span from Hicks Street to Clinton Street along Atlantic Avenue. The store front would contain artifacts from the tunnel as well as continue the the tunnels time traveling motif. The store front would lead to a stairway which would take visitors to various sub basements in the building. It is widely believed many of the old buildings boring the tunnel have basements at the the tunnel level and may even have existing connections into the tunnel. News paper articles from the era state that attempts were made to link the buildings to the tunnel. Additionally, a station accessible by storefront is not a very fetched idea for the time period. In 1870, Alfred E. Beach opened New York City's first subway line the Beach Pneumatic Subway under Broadway near City Hall using this paradigm. An architectural team would have to determine the validity of these claims since building records from the 1840s do not appear to be extant.

All engineering assessments will be made by professionals who will gain tunnel access through the approval of a relevant city agency. Following this approval a maintenance of traffic plan will be submitted to the city for the protection of those making improvements in the tunnel.

We anticipate the implementation of this plan will make the entire tunnel fire proof, and that Emergency Personnel entering the tunnel on a job will need to carry only a minimum of appropriate equipment, and traditional "gurneys" will easily fit within the tunnel.

The tunnel entertainment experience will begin to evolve during phase 2. Audio visual equipment will begin to be installed in order to create the immersive virtual experience described in the above chapters. This will be purchased with BHRA's visitor revenue as well as money from corporate sponsorship. The multimedia equipment may cost up to $5 million and will put the tunnel on par with off broadway shows. The use of multimedia equipment will make the tunnel a truly unique experience that can not be found else where in the city.
Below: Empty Store Fronts on Atlantic Avenue.

Above: Empty Store Fronts Along Atlantic Avenue   *Below*: Drawing 1/3 of New Entrances

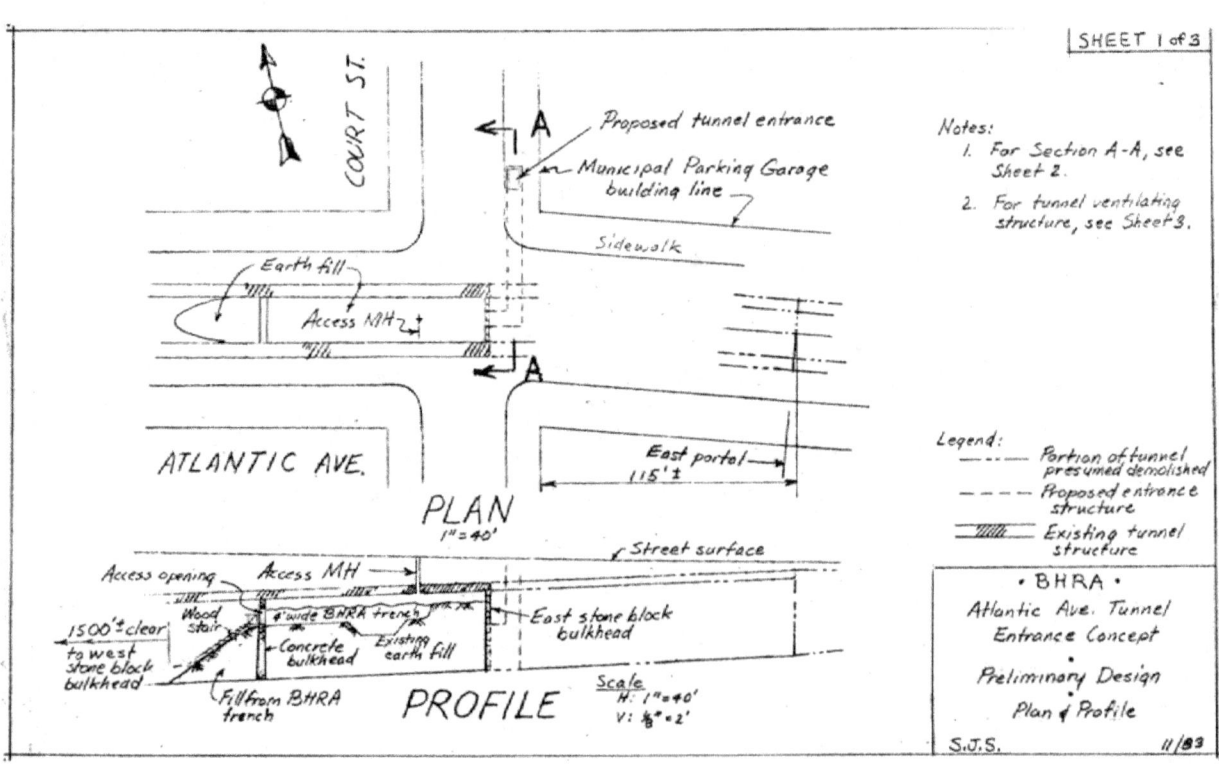

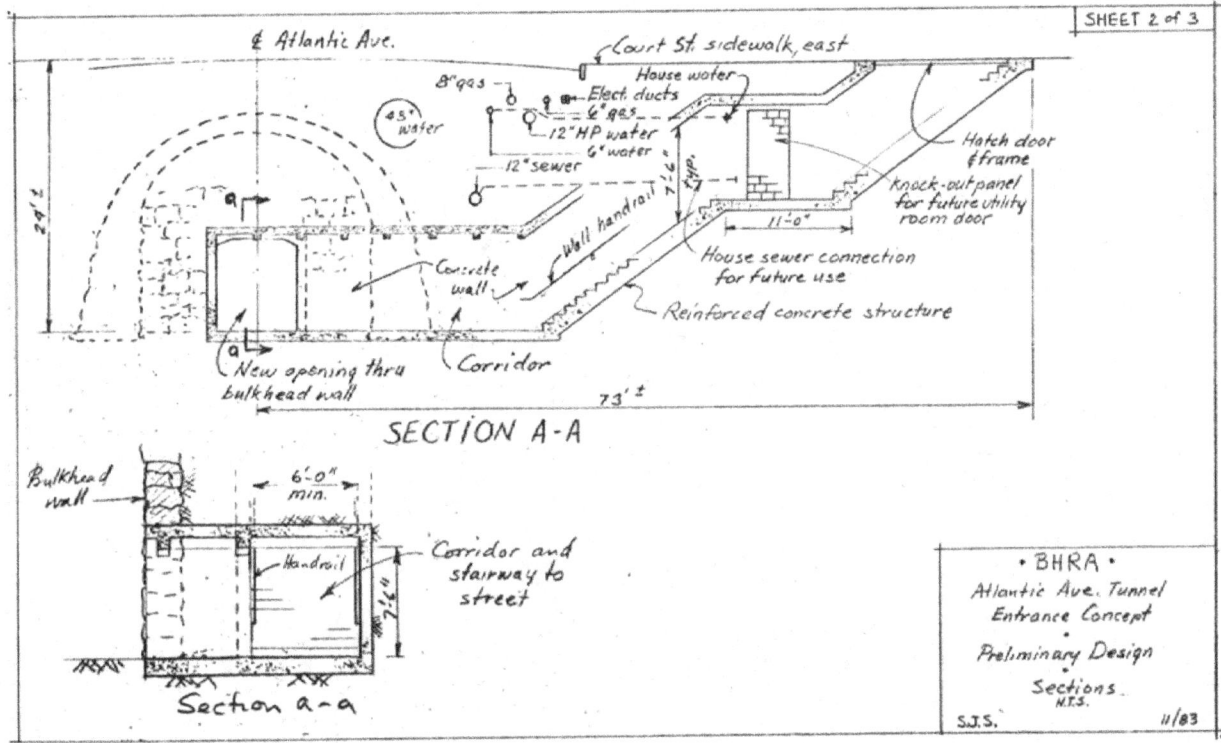

Above: Drawing 2/3 of New Entrance   Below Left: Historically Themed Elevator   Below Right: Period Newspaper Article Describing A Proposed Tunnel Access Via An Existing Adjacent Building's Basement.

COMMON COUNCIL COMMITTEES.

Meeting of the Grading and Paving, Railroad and Special Investigating Committees—The Atlantic Street Tunnel—Nicolson Pavement Job—Southside Railroad—The Lost Petitions—Important Matter to Come Before the Ferry and Water Rights Committes.

The Grading and Paving Committee of the Common Council met last evening at the City Hall, Ald. O'Keeffe in the chair.

The petition of J. P. Stryker and others, dated Sept. 10th, asking to be granted a lease of the tunnel under Atlantic street, between Hicks and Court streets, was taken up. The petitioners asked for the privilege of cutting entrances from property at or above Clinton street, under the sidewalk, and gas, water and sewer pipes, to the tunnel, and to partition the same with walls, for the purpose of general storage, excepting explosive articles.

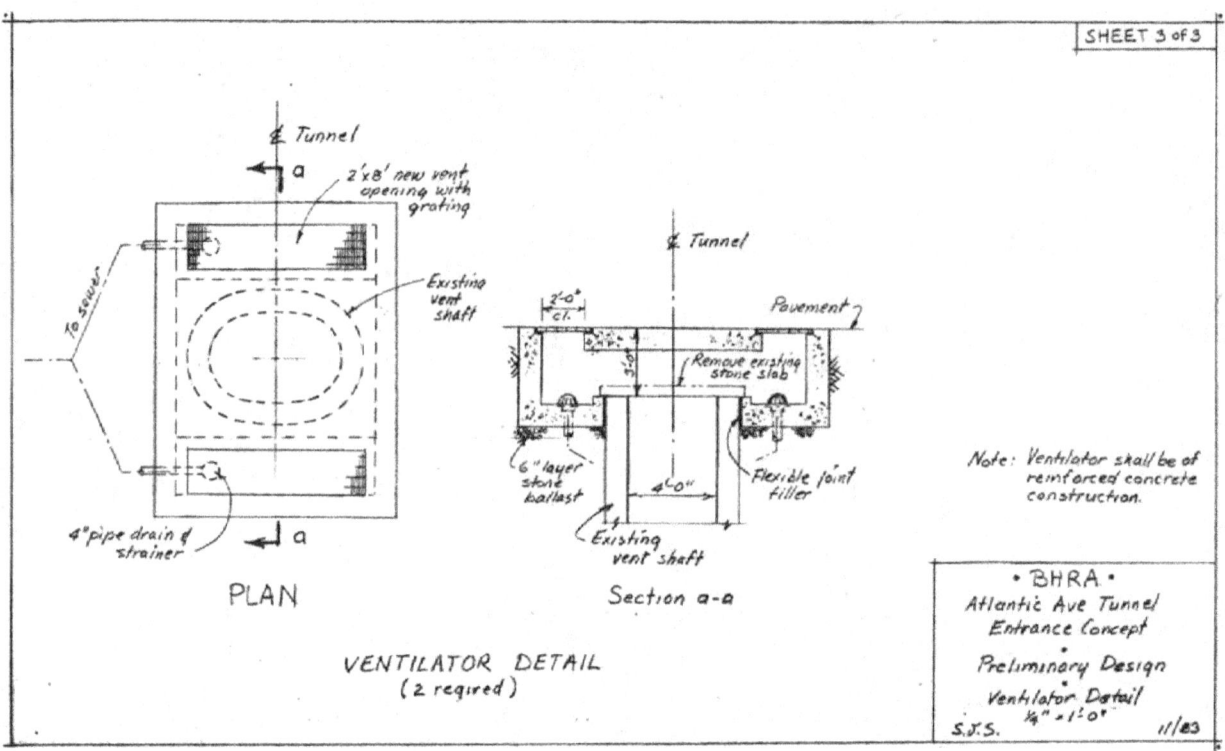

Above: Drawing 3/3 Tunnel Ventilators

Opportunities For Sidewalk Access:
Based solely upon the criteria of minimizing the relocation of existing utilities, we suggest the S/E corner of Atlantic Avenue and Court Street, and/or the S/E corner of Atlantic Avenue and Hicks Street.

Opportunity For Elevator Access:
Based solely upon the criteria of minimizing the relocation of existing utilities, we suggest the S/E corner of Atlantic Avenue and Hicks Street.

Opportunity For Building Basement Access:
1. The vicinity of 71 Atlantic Avenue: this row of buildings line up directly with the location of the presumed buried locomotive.

2. The vicinity of 110 Atlantic Avenue: this row of structures could provide access to the mid- section of the tunnel.

3. The vicinity of 156 Atlantic Avenue: this particular building has three sub-basements, the lowest of which lines up with track level inside the tunnel. Other buildings in this row may also have triple basements, offering the opportunity for a visitor orientation center, rest rooms, etc outside of the tunnel structure itself.

THE LOCOMOTIVE

According to historical records and sensor data there is a wood burning steam locomotive from the mid 1800s buried in the sealed up portion of the Atlantic Avenue Tunnel. It was not uncommon for obsolete technology (old trains, cars, etc) to used for fill in the 19th and early part of the 20th centuries. BHRA proposes a series of test borings accessing the artifact for further study, and then direct observation of the artifact by use of an excavated pit or trench.

The team will first remove the existing street pavement and subsurface layers of construction until the top of the tunnel crown is encountered. A hammer will breach the tunnel crown and a backhoe will subsequently remove the pavement and fill material down to within approximately five feet of the artifact's expected height. An engineered shoring system will be installed as the depth of the excavated trench increases. The team will then hand excavate down to the artifact level and shoring will continuously be installed as work progresses deeper into the tunnel. Ideally, the trench will be offset from the artifact's centerline so as to expose its side. The project team has compiled a proposed borehole plan based upon the data collected during the geophysical survey and utility mapping phases in order to avoid

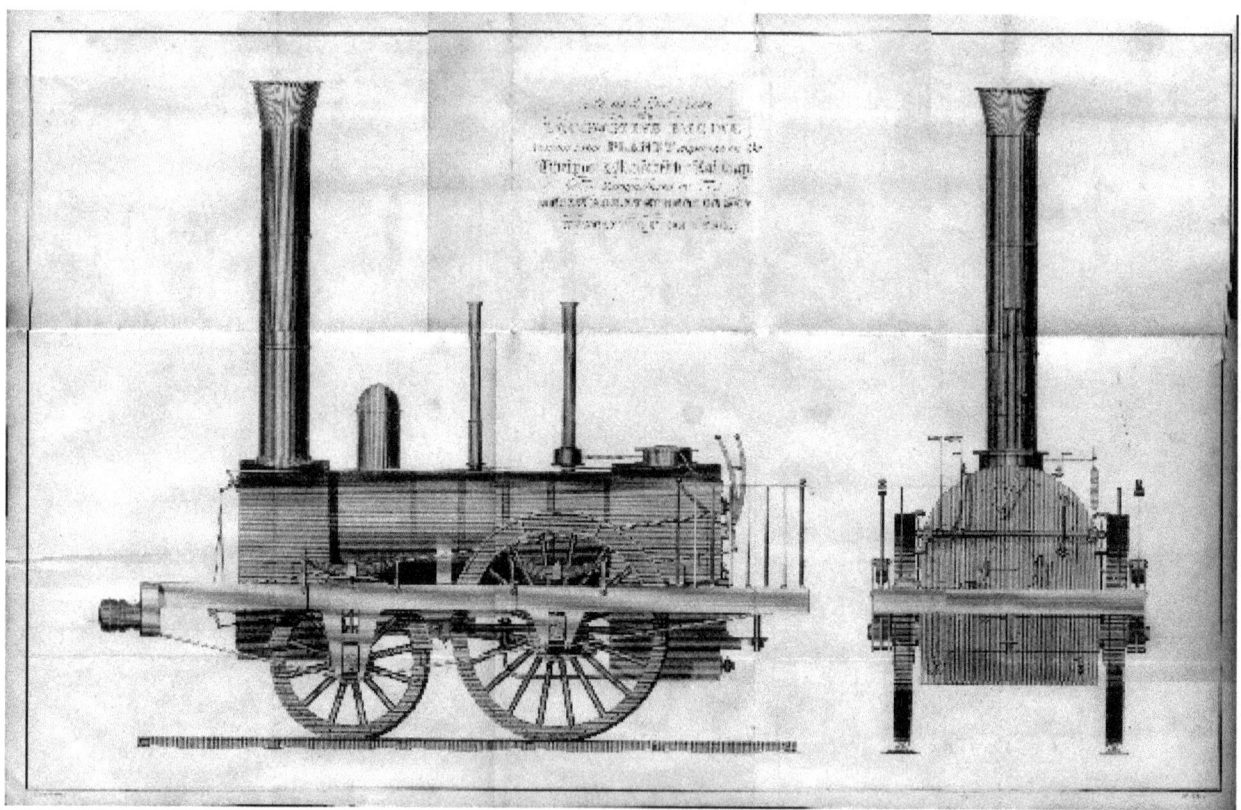

interference with existing below grade utilities. The team will coordinate the proposed drilling schedule with the traffic engineer in order to prepare a phased traffic control plan in sequence with the borehole schedule.

All documentation will be submitted to the City for review and approval prior to submitting a permit application with the NYCDOT Permit Office. The necessary permits to facilitate the proposed work include Permit #0211 for a Temporary Closing of Roadway, and Permit #0126 for Test Pits, Cores or Boring. The team will apply for both permits to execute the proposed work.

Prior to commencement of work, the team will coordinate a One-Call utility mark out through Dig Safely New York in order to physically mark the pavement along Atlantic Avenue and avoid any active utilities during drilling. The team will provide and operate a heavy-duty Vactor and direct push Geoprobe to install up to nine (9) borings within and around the location where the geophysical anomaly has been identified.

The borings will be advanced via four (4) inch diameter boreholes. It is estimated that the team can complete three locations per day. Once the boreholes are advanced to the desired depth, the team will install a color video inspection camera to inspect and photo document the area below the boreholes for evidence of the large metallic anomaly.

The objective is to further delineate the anomaly's location and orientation as well as attempt to visually inspect the target object in order to ascertain its identity. Surface utilities, buried debris in historic fill, and other obstructions may inhibit the advancement of borings in select locations.

For this reason, a per day rig charge will apply. Ideally, a portion of the artifact will be encountered within one of the borings. The borings will also provide information on the location of voids as well as depth and content of fill material. Two samples of the excavated fill will to be lab tested for soil composition (Sieve Analysis) as well as for contaminants (TCL/ TALs) to determine if there are any environmental concerns for the workers and the potential costs related to the disposal of contaminated soil.

The contents of the fill material should also be inspected and analyzed by the team's archaeologist and conservator. Any artifact retained will be processed, identified, and catalogued. Upon completion of work, drillers are to backfill cuttings in boreholes, tamp the backfill material, and return the roadway surface to its original condition per NYCDOT requirements and standards

In the event that the boreholes demonstrate further evidence of the anomaly and camera inspection of the artifact verifies the existence of a buried metallic object, the team will identify a definitive location for invasive exploration. The optimal location will also avoid interference with existing, active utilities.

The team will prepare a site plan for the proposed trench, which will be used as the basis for permitting, and traffic control planning.

The team will also design the necessary shoring for excavation. The trench will be large enough to enable a two-person team to inspect and work within the available space. Excavation through the street level and tunnel fill will require the design of a vertical shoring system, use of an operated backhoe, as well as additional archaeological consultant time in order to screen soils and document the trench. The trench will be approximately 3'-0" wide, 5'-0" long, and 20'-0" deep and it is estimated that it would require two days to complete the trench excavation.

We have allocated time for a conservator to inspect the artifact in situ, observe the surroundings, and develop a stabilization and conservation plan. The locomotive must be stabilized in place prior to completing the work. Upon completion of work, the artifact will be covered with a

specified protective geotextile fabric, the trenches filled in with clean granular and compacted fill, and the street repaired per NYCDOT standard.
Any artifact of record would become property of the people of the City of New York. BHRA would act as custodians and interpret the significance artifacts at the tunnel site.

BHRA proposes that it solicits contractors and with DOT, oversees work for improved access to the tunnel as well as work for recovery of steam locomotive. With the City providing funding as a public private partnership BHRA believes as a non-profit, it can get the work done faster, and at lower cost.

In 1986, a group of volunteers in England began planning for the construction of a replica Planet type steam locomotive. This is the type of locomotive that is believed to be buried under Atlantic Avenue. If recovered this locomotive would be the last original of its kind. The replica is a working steam model constructed with modern building techniques and has travelled over 1000 miles since its completion in 1992. The replica currently resides at the Museum of Science and Industry in Manchester. The BHRA proposes that a similar replica be constructed for the entertainment and education of visitors at the tunnel. Visitors will get a historically accurate experience of the tunnel at the time of its functioning.

PROJECT BUDGET

Summary of Work

Construction of one or more subway-style sidewalk entrances to the tunnel at Court Street and Atlantic Avenue, as per attached drawings. A kiosk will also be built to protect the entrance and provide shelter. One or more existing ventilation shafts will be reopened and activated to provide forced-air ventilation. Also to be installed are standpipes for fire protection and an improved museum-style lighting system. Construction of a museum within the tunnel. Exhibits will highlight the impact the tunnel had on the economic and social development (Transit Oriented Development) of Brooklyn; in particular, the reason it was built, how it was built and why it was closed. Various eclectic scientific and historical principles, and cutting edge theories relating to rail transit science and local history will also be included in the educational presentation within the tunnel. The centerpiece of the museum will be the circa-1830's steam locomotive discovered in the tunnel. A method was devised to drive an approximately 60 foot long "connecting tunnel" between the buried locomotive, and the interior of the tunnel under Atlantic Avenue and Hicks Street. This work is anticipated to be accomplished without any contact with existing underground utilities, through the use of pure tunneling, and by not employing the "cut and cover method" at all. Other exhibits will include artifacts from the tunnel and various media illustrating the tunnel in use, and Brooklyn in general during that period of time. Another major attraction will be a replica locomotive.

Description	Cost
Court Street Entrance	$2,661,569
Hicks Street Entance w/Elevator	$7,000,000
Discovery of Lost Engine	$150,000
Safety Upgrades (Exhaust Fans, Fire Stand Pipes, & AED)	$154,577
Telecommunications Connectivity	$2,217
Plumbing	$62,103
Electrical	$106,463
Total	**$ 10,136,929**

UPDATED COST ESTIMATES

Original costs were estimated by Stephen L. Carroll P.E. in 1985

Updated Cost Estimates Using Engineering News Record Cost Index:

1985 Cost Estimates for 1 Tunnel Entrance: $1,200,000
1985 Construction Cost Index[1]: 5388.08
July 2015 Construction Cost Index[2]: 10037.40

Inflation Factor from 1985 to 2015

10037.40/5388.08 = 1.86

Updated 2015 Cost

1,200,000 x 1.86 = $2,235,412

Cost Per Each Entrance: $2,235,412

Updated Cost Estimate Using Consumer Price Index (CPI)

Cumulative Rate of Inflation: 121.8%[3]

1985 Cost: $1,200,000

2015 Cost: $2,661,568.77

---

[1] From Engineering News Record

[2] From Engineering News Record

[3] http://www.usinflationcalculator.com/

TIME LINE OF EVENTS

1979-1980 - Robert Diamond learns about the tunnel and begins investigating its whereabouts

1981 - Robert Diamond discovers the Atlantic Avenue Tunnel

1982 - Tunnel access improved by Robert Diamond & friends

1982 - BHRA formed

1989 - Tunnel gets designation by National Register of Historic Places

1989 - Streetcar Project begins in DUMBO, Brooklyn.

1994 - The first streetcar in decades runs under its own power in Brooklyn

1996 - Red Hook incorporates the streetcar into its 197a plan for regeneration

1996 - BHRA gets federal ISTEA grant funds to begin construction on the Red Hook streetcar line

1999 - Construction begins on the streetcar line

2001- NYC DOT approves construction of streetcar onto public streets

2001 - 12 PCC streetcars purchased from Buffalo, NY

2003 - Streetcar project halted

2005 - Streetcars and BHRA equipment stolen

2005 - Federal grant secured for a streetcar study by the NYC DOT

2010 - Tunnel tours halted

2010 - Scanning devices detect the locomotive long believed to be buried in the tunnel

2011 - NYC DOT completes streetcar study

2011 - Guinness Book of World Records dubs the Atlantic Avenue Tunnel worlds oldest subway tunnel

LIST OF SUPPORTERS

- Former Mayor Ed Koch

- Jack Lusk - Former Special Advisor to Mayor Koch & Director of the Mayor's Transportation Office

- Howard Golden - Former Brooklyn Borough President

- Harrison J. Goldin - Former NYC Comptroller.

- Abe Gerges- Former NYC Council Member

- Jan C. Childress - Executive assistant to Former U.S. Rep. Freddy Richmond

- John Herzog - Historian, specializing in Scripophily.

- Gerard Sofian, P.E.- Former NYCDOT Assistant Commissioner

- Paul Stanton - Former NYCDOT Agency Chief Contracting Officer

- Ross Johnson - Former supervisor at MTA NYC Subways Third Rail Dept.

- Bill Rhodes - Former supervisor at MTA NYC Subways Third Rail Dept.

- Jon Graham - Former NYCDOT Director of Franchises and Consents

- Peter Dunleavy, R.A. - NY State DOT

- Anthony S. Cosentino, P.E. - Former NYCDOT Chief of Builder's Pavement Bureau & Chief Engineer of NYCDEP.

- Alan L. Smith- Former Assistant VP, Brooklyn Union Gas

- Former NY Governor Nelson Rockefeller

- Luke Stets, Technologist & Government Publicist

- Stephen Levin - NYC Council Member District 33

- Peter Yost - Head of Pangloss Films.

- Sam "Gridlock Sam" Schwartz, P.E. - transportation engineer & Former NYCDOT Deputy Commissioner

- Roy Sloan - President of the Cobble Hill Association

- Sandy Balboza - President of the Atlantic Avenue BID,

- Tony Giordano - President, Sunset Park BID,
- Andrew Costello - Director of NYPD Brooklyn Transit Bureau
- John Quadrozzi Jr.- Owner, Gowanus Bay Terminal
- Greg O'Connell Sr. - O'Connell Real Estate Organization

599 E. 7TH ST., SUITE 5A BROOKLYN, N.Y. 11218   WWW.BROOKLYNRAIL.NET

PRESS COVERAGE

The New York Times

Brooklyn Daily Eagle

DAILY NEWS

DNAinfo.com New York neighborhood news

Newsweek

599 E. 7TH ST., SUITE 5A BROOKLYN, N.Y. 11218   WWW.BROOKLYNRAIL.NET

BROOKLYN HISTORIC RAILWAY ASSOCIATION

599 EAST **7**TH STREET, SUITE **5**A

BROOKLYN, N.Y. **11218**

APPENDIX

- LMW Engineering Group LLC - Structural Integrity Evaluation
- Singstad, Hurka, & Associates - Structural Integrity Evaluation
- City of New York Department of General Services Laboratory - Core Samples
- MTA - Live Loads Data
- DEP - Bureau of Water Supply
- FDNY - Safety Requirements
- DEP - Bureau of Sewers
- City of New York Department of General Services - Division of Public Structures
- DOT - Office of the Commissioner
- Steven L. Carroll P.E. - Tunnel Entrance Cost Estimate
- Updated Costs for Inflation
- **Methods For Inflating Construction Costs to 2015 Dollars**
- Brinkerhoff Environmental Services - Atlantic Avenue Tunnel Investigation Project
- ZOLL Medical Corporation - AED
- DDC - Atlantic Avenue Underground Utilities
- Estimated Elevator Cost - MTA Paid $7 Million

Engineering Reports and Studies

LMW Engineering Group, LLC
2539 Brunswick Ave. Linden, NJ, 07036
WWW.LMW-ENG.COM
Tel.(908) 862-7600 Fax(908) 862-8998

March 25, 2009

BROOKLYN HISTORIC RAILWAY ASSOCIATION
599 E 7th Street, Suite 5A
Brooklyn, NY 11218

ATTN.: Robert Diamond, President

RE: Atlantic Avenue Tunnel,
between Boerum Place and Columbia Street

SUBJECT: Structural Integrity Evaluation

Dear Mr. Diamond;

Pursuant to your request, we have performed an evaluation of the structural integrity of the referenced tunnel structure, based on visual inspection tour of the tunnel site, and review of reports of previews evaluations performed by others. This letter-report presents the results of our evaluation.

The inspection tour of the tunnel was performed by our senior engineer, Mr. Tony Onyeagoro, P.E., in the afternoon hours of Wednesday, March 25, 2009, assisted by one of your associates. Access and egress to the site were through a manhole located at the middle of the roadway, intersection of Atlantic Avenue and Court Street, Borough of Brooklyn. Visibility in the tunnel was generally subdued, but details of the structural elements were clearly observable using a flash light. Select, typical portions of the structure details were captured using flash-enabled *Canon PowerShot A540* camera.

The tunnel structure consists of a masonry block arched dome upper section, supported on either side by a wall made of varying-size stone or rock quarry, embedded in a matrix of very strong grout mix. The composition of the structural elements is relatively consistent for the entire length of the tunnel. Evidence of removed rail lines and ballasts are observable through out the general grade of the tunnel, which is firm with no sign of moisture or subsidence. The general condition of the tunnel structural elements, relative to previous reports, is impressively devoid of any sign of deterioration. All masonry block units appear to be intact, with no visible loosening of joints. Similarly, there was no observable loss of filler materials within the walls' stone matrices.

Accordingly, in concurrence with the findings of previous evaluations, please be advised that:

1. The tunnel structural members were inspected by our engineer.
2. The load carrying capacity of the tunnel is sufficient to support the anticipated loading (overburden and live load).
3. The non-load carrying members are secure.
4. The structural integrity of the tunnel is sound and has not been compromised by aging.

Certified Minority Business Enterprise NY/NJ
Engineering Consultation ● Design ● Inspection ● Testing

5. The tunnel can be considered safe under its current use for visitors and tourist attraction.
6. There is no evidence that any form of maintenance or repair work is necessary at this stage.
7. The tunnel can be safely, with relatively minimum rehabilitation effort, mostly esthetic, be utilized as a museum or similar facility.
8. In summary, the tunnel, as inspected by us, is a safe and sound structure.

We appreciate the opportunity to provide this service for you, and look forward to continuing involvement in this extraordinarily noble venture. If you have any questions regarding this letter-report, please feel free to call us.

Very truly yours,
LMW ENGINEERING GROUP, LLC

Tony C. Onyeagoro, P.E.
Senior Project Engineer
NYS P.E. License # 063593

SINGSTAD, HURKA & ASSOCIATES, P.C.
Consulting Engineers

Frank P. Hurka, P.E.
Anthony S. Caserta, P.E.
Joseph A. Defino, R.A.
Stanislaw Szatowski, P.E.
Simon Zabrocky, P.E.

November 7, 1985

Brooklyn Historic Railway Association
599 E. 7th Street
Brooklyn, NY 11218

      Re: Atlantic Avenue Tunnel
           Engineering Report

Gentlemen:

    In accordance with your request, I have completed the engineering analysis of the structural integrity of "The Atlantic Avenue Tunnel". The results of my investigation and the calculations are incorporated in the attached report entitled "Engineering Report - Atlantic Avenue Tunnel". This report is supplementary to my interim report of said structure dated August 7, 1984. (See Appendix A).

    I have concluded that the existing tunnel structure is a safe and sound structure and that its structural integrity has not been compromised with age. The masonry and stone tunnel can sustain its present overburden load, and the live loads on Atlantic Avenue.

    In my opinion, the useful life of the tunnel is not in jeopardy and the facility can be used to house an underground museum or exhibit hall.

    Kindly advise if there are any questions or if I can be of any further assistance.

                                      Respectfully submitted,

                                      Anthony S. Caserta, P.E.

Attachments

SINGSTAD, HURKA & ASSOCIATES, P.C.

ENGINEERING REPORT
ATLANTIC AVENUE TUNNEL

Prepared by:

A.S. Caserta
P.E. No. 34498
NY State

SINGSTAD, HURKA & ASSOCIATES, P.C.

I. HISTORY OF THE TUNNEL

Construction on the railroad horse-shoe shaped tunnel started in May 1844 and was completed seven months later in December 1844. The entire structure was built by several hundred laborers using only hand tools such as pick, shovels and pack mules. The physical dimensions of the tunnel are impressive even by modern standards. The horseshoe shaped tunnel is twenty one (21) feet wide, seventeen (17) feet six (6) inches high and one thousand nine hundred and fifteen (1915) feet long. The tunnel has a brick arch roof section varying in thickness from 20 inches at the crown to 4 feet at the spring line. The stone masonry walls supporting the arch vary from 4 feet (spring line) to 6 feet at the base. (see Fig. 1).

The structure was built by the "cut and cover" method. Simply stated, this construction technique involves excavating from the surface grade down to the tunnel invert, erecting the side walls and roof arch, and then backfilling and repaving.

The tunnel was ventilated by the use of 3 large ventilation shafts extending to the street above. These ducts are oval-shaped with a maximum width of approximately 6 feet and are spaced 325 feet on centers along the middle third of the tunnel.

It is estimated that the tunnel was sealed up permanently in 1861 by the erection of 2 stone masonry walls inside of each portal; reducing the tunnel length to 1500 feet

II. TEST CORE SAMPLES AND TESTING

To complete the interim report dated August 7, 1984, the Brooklyn Historic Railway Association arranged to take core samples of the brick, stone and soil of the tunnel structure and have them tested.

The locations of the core samples were established by Mr. Caserta and selected to provide representative samples of the tunnel structure. (See Fig. 2). The test cores served two purposes as follows:

- Verify the thickness of the tunnel arch and wall sections

- Obtain compressive strength (p.s.i.) values of brick and stone by standard laboratory testing procedures

- Determine grain size analysis, sieve analysis and characteristics of existing soil in the invert and behind the tunnel structures.

The core drilling operation was done by the Semcor Inc. and completed on June 11, 1985.

SINGSTAD, HURKA & ASSOCIATES, P.C.

The core samples were deliever to the New York City Department of General Services Laboratory in July 1985. The tests outlined above were performed by the lab and completed on August 5, 1985 (See Appendix B).

III. STRUCTURAL ADEQUACY OF TUNNEL

As inspection was made of the tunnel by Mr. A. S. Caserta on July 19, 1984 and again on July 8, 1985. At both inspections, the masonry tunnel lining was observed to be:

- Free of any deterioration
- Showed no signs of distress
- Showed no signs of distortion
- Free of lossen brick or mortar

There was no presence of any water inflow at the invert or evidence of water infiltration through the masonary walls.

The laboratory report (NYC Dept. of General Services Laboratory) states that the compressive strength tests on the core samples yield an average value of 3151 p.s.i. (See Appendix B). Assuming an allowable compressive stress of 500 psi (0.15 f'c), it is apparent that the Atlantic Avenue Tunnel is satisfactory under the conservative overburden loading shown in the stress calculation.

The tunnel structural investigative analysis is based on several assumptions as follows:

1. The masonry tunnel lining has the lateral support of the surrounding sandy soil.

2. The tunnel lining thickness as determined by the cores is consistent for the entire tunnel length.

3. Since masonry cannot take tensile stress, the thrust (T) load is distributed over a smaller area.

The design overburden load on the lining is taken as the average depth of soil cover over the tunnel. Additionally, a live load of 500 p.s.f. for street traffic is superimposed in the overburden. For a design load of 10 feet of earth and a 500 p.s.f. live load, the calculations indicate a positive moment of 14,210 ft.lbs with a compressive thrust of 21,000 lbs per foot of tunnel at the spring line, Therefore the compressive stress in the masonry lining is approximately 390 p.s.i.

SINGSTAD, HURKA & ASSOCIATES, P.C.

3.

The vertical side stone walls of the tunnel are bearing on non-cohesive soil and exerts a bearing load of 7300 p.s.f. at its foundation. The soil samples taken in the invert were tested and exhibit the characteristics of a granular soil without any clay content. According to the text "Foundation of Structures - Dunham, 3rd Edition", a conservative allowable bearing value for this material is 6000 to 8000 p.s.f.

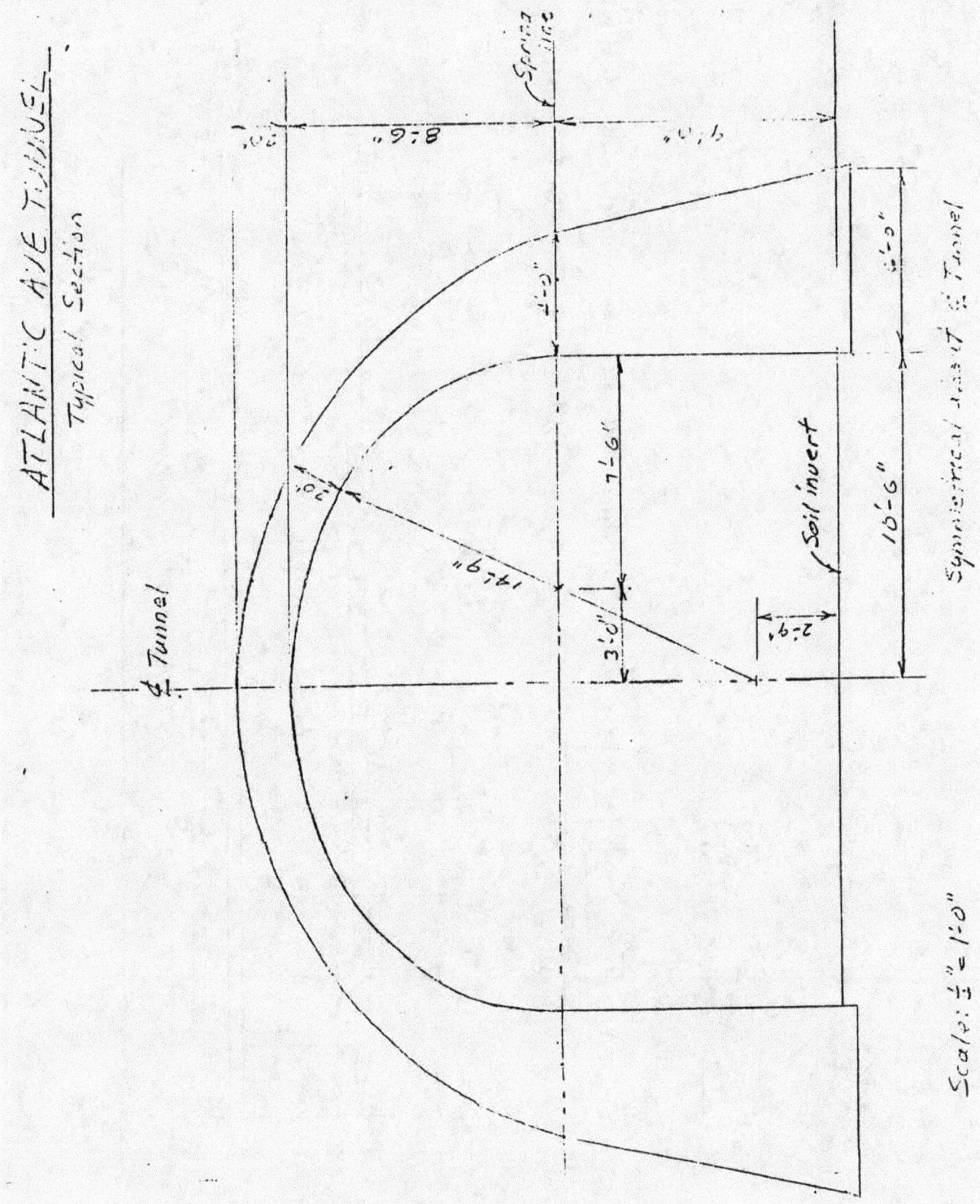

FIG. 1

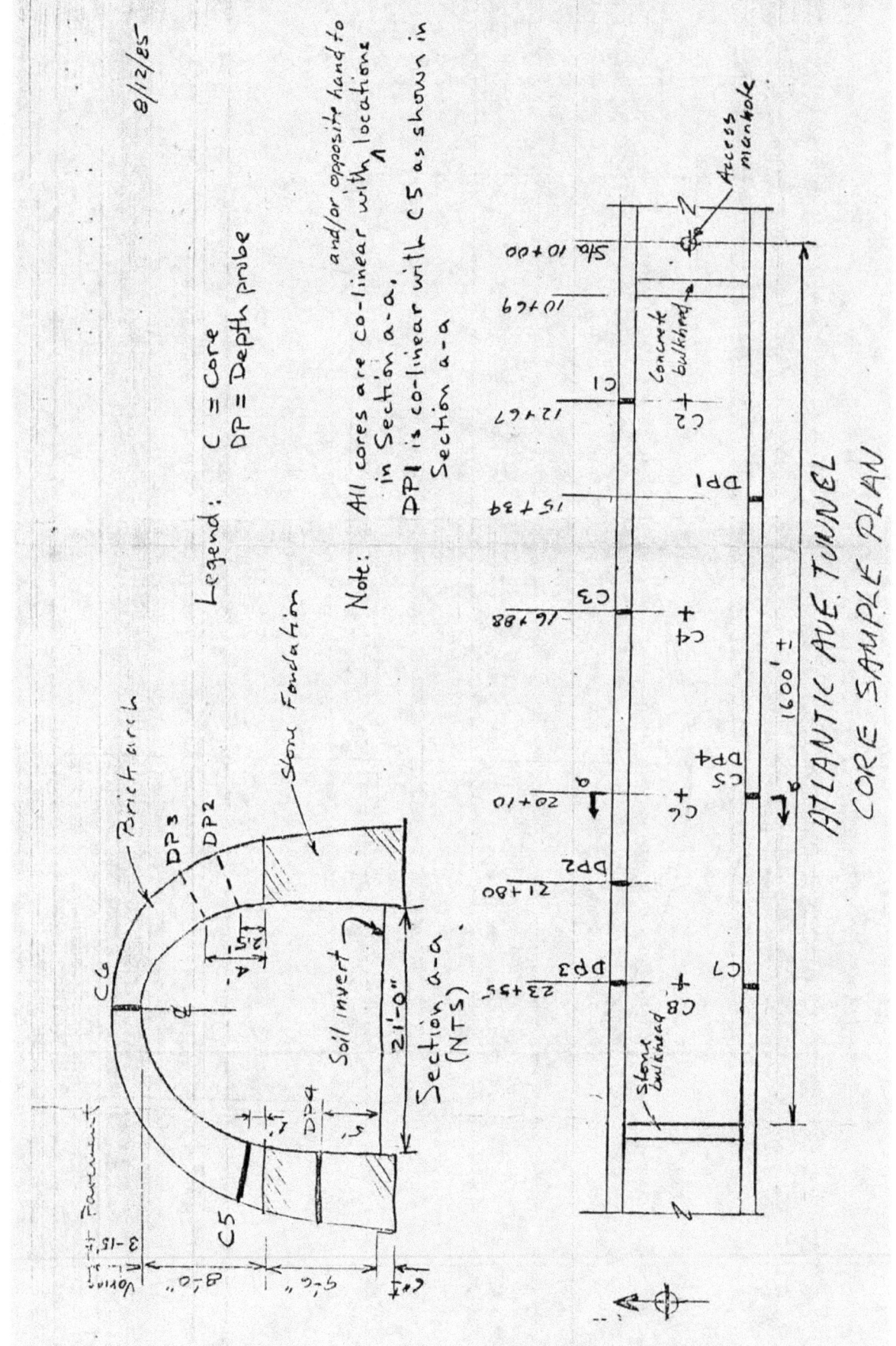

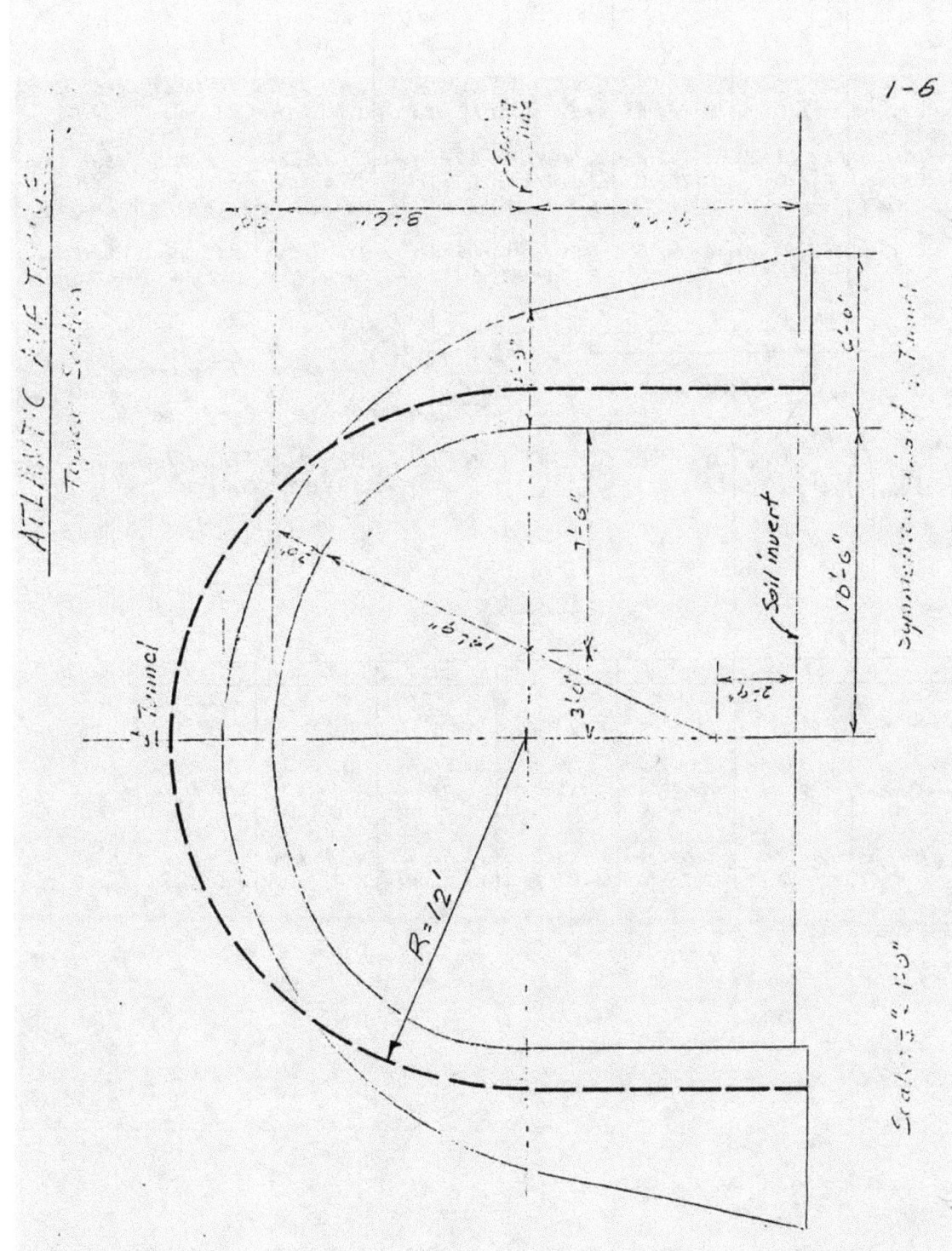

SINGSTAD, KEHART, NOVEMBER, & HURKA
CONSULTING ENGINEERS

PROJECT: Atlantic Ave Tunnel

$$M\phi = -H(h + r\sin\phi) - Pr\sin\phi + \frac{wr^2}{2}\sin^2\phi$$

$H = -0.07123\, wr$
$P = 0.56769\, wr$

	0° S.L.	15°	30°	45°	60°	75°	90° Crown
$\sin\phi$	0	0.25882	0.50	.70711	.86603	.96593	1.0
$\sin^2\phi$	0	0.06699	0.25	.500	.75000	.93302	1.0
$r\sin\phi$	0	3.10584	6.00	8.48528	10.39236	11.59116	12.0
$r\sin^2\phi$	0	0.80388	3.00	6.00	9.00000	11.19624	12.0
$h + r\sin\phi$	9.5	12.10584	15.50	17.98528	19.89236	21.09116	21.5
$-H(h+r\sin\phi)$	+0.67669 wr	+0.86230 wr	+1.10407 wr	+1.28109 wr	+1.41693 wr	+1.50232 wr	+1.53145 wr
$-Pr\sin\phi$	−0	−1.76315	−3.40614	−4.81701	−5.89964	−6.58019	−6.81228
$+.5r\sin^2\phi$	+0	+0.40194	+1.5000	+3.00	+4.5000	+5.59812	+6.00000
$M\phi = \{$	+0.67669 wr	−0.49891 wr	−0.80207 wr	−0.53592 wr	+0.01729 wr	+0.52025 wr	+0.71917 wr
	+8.12028 w	−5.98692 w	−9.62484 w	−6.43104 w	+0.20748 w	+6.24300 w	+8.63004 w

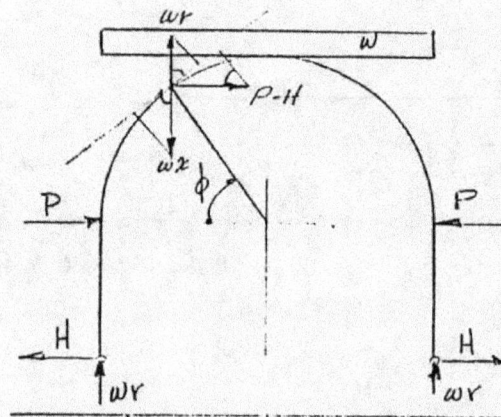

$$T = (wr - wx)\cos\phi + (P - H)\sin\phi$$
$$T = wr\cos^2\phi + (P - H)\sin\phi$$
$$(P - H) = (0.56769 - 0.07123) wr = 0.49646\, wr$$

	0° S.L.	15°	30°	45°	60°	75°	90° CROWN
$\sin\phi$	0	.25882	0.50	.70711	.86603	.96593	1.0
$\cos^2\phi$	1.0	0.93301	.75	.50	.25	.06699	0
$wr\cos^2\phi$	1.0 wr	.93301 wr	.75 wr	.50 wr	.25 wr	.06699 wr	0
$.49646\, wr\sin\phi$	0	.12849 wr	.24823 wr	.35105 wr	.42995 wr	.47955 wr	.49646 wr
THRUST = {	1.0 wr	1.06150 wr	0.99823 wr	.85105 wr	.67995 wr	.54654 wr	.49646 wr
	12 w	12.738 w	11.97876 w	10.21260 w	8.15940 w	6.55848 w	5.95752 w

Project: Atlantic Ave Tunnel
Sheet Number: 3 of 6

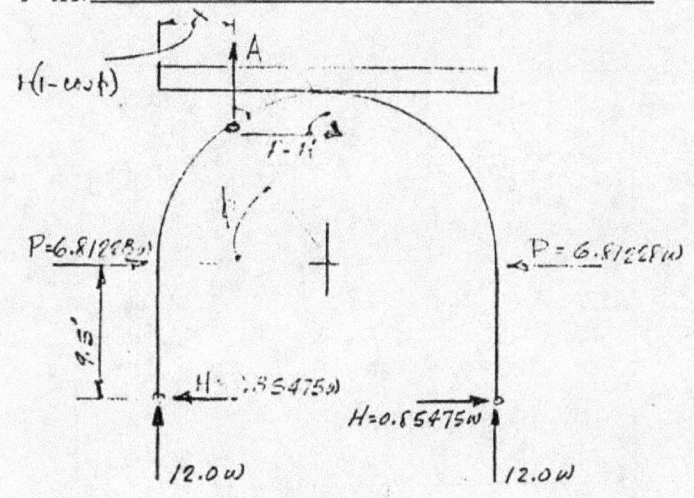

$1 - H = 0.4961... \, wr$
$\quad\quad = 5.95752 \, w$

$A = 12w - wr(1-\cos\phi)$

$V = (P-H)\cos\phi - A\sin\phi$

	0° S.L	15	30	45	60	75	90 Cr
$\cos\phi$	1.0	.96593	.86603	.70711	.50	.25882	0
$\sin\phi$	0	.25882	.50	.70711	.86603	.96593	1.0
$r(1-\cos\phi)$	0	.40884	1.60764	3.51468	6.0	8.89416	12.0
A	12w	11.59116w	10.39236w	8.48532w	6.0w	3.10584w	0
$(P-H)\cos\phi$	5.9575w	5.75455w	5.15939w	4.21262w	2.97876w	1.54193w	0
$A\sin\phi$	0	3.00002w	5.19618w	6.00005w	5.19618w	3.00002w	0
$V\phi$	5.93752w	2.75453w	-0.03679w	-1.78743w	-2.21742w	-1.45809w	0

Uniform Vertical Load

@ 4' × 125 #/ft ; $\omega = 500$ #/ft
@ 6' × 125 #/ft ; $\omega = 750$ #/ft
@ 10' × 125 #/ft ; $\omega = 1250$ #/ft

SINGSTAD, KEHART, NOVEMBER, & HURKA
CONSULTING ENGINEERS

PROJECT: Atlantic Ave. Tunnel

SPRING LINE

	MOMENT (FT LBS)				THRUST (LBS)			
	4'	6'	10'	15'	4'	6'	10'	15'
Uniform load	+4060	6090	10,150	15,225	6000	9000	15000	22,500

@ 30° FROM S.L

	MOMENT (FT. LBS)				THRUST (LBS)			
	4'	6'	10'	15'	4'	6'	10'	15'
Uniform Load	-4812	-7219	-12031	-18047	5989	8984	14,973	22,460

@ CROWN

	MOMENT (FT. LBS)				THRUST (LBS)			
	4'	6'	10'	15'	4'	6'	10'	15'
Uniform Load	+4315	6473	10,788	16,181	2979	4468	7447	11,170

LIVE LOAD

$W_{LL} = 500 \text{ psf}$ is equivalent to 4' cover @ 125 #/ft³

∴ For Investigation of Stresses — use average 10' + L.L.

Crown: $M_{@10'} = 10,788$ ft-lbs $T_{@10'} = 7447$ lbs
$\underline{M_{LL} = 4315 \text{ ft-lbs}}$ $\underline{T_{LL} = 2979 \text{ lbs}}$
15,103 ft-lbs   10,426 lbs

PROJECT: ATLANTIC AVE. TUNNEL

@ 30° From ?.?.

M_∂ — -12,031 $T_{@10'} = 14,973$

M_{LL} — -4812 $T_{LL} = 5987$

-16,843 ft-lbs   20,962 lbs

Spring Line

$M_{@10'} = +10,150$ $T_{@10'}$ 15,000

$M_{LL} = +4060$ $T_{LL} = 6,000$

14,210 ft-lbs   21,000 lbs

STRESS CALCULATIONS

$$f_c = \frac{T}{A} \pm \frac{6M}{bd^2} \; ; \; f_{c\,max} = \frac{2T}{3bq} \text{ if tensile stress exists}$$

15' Cover w/LL	M ft-lbs	T lbs	b	d	d^2	f_c (psi)	f_t (psi)	q	$f_{c\,max}$ (psi)
Sp.line	14,210	21,000	12	20	400	87.5	213.2	3	388.7
30°	-16,843	20,962	12	20	400	87.3	252.6	3	388.2
Crown	15,103	10,426	12	20	400	43.4	226.5	3	193.1

BEARING VALUES

$33' \cdot 10' (avg) \times 125 \text{ pcf} = 41,250 \text{ lbs/ft}$

$\frac{20'}{2} \cdot 24' \cdot 2.83' \cdot 150 \text{ pcf} = 10,195$

$6 \times 15 \cdot 8' (avg) \times 120 \text{ pcf} = 14,400$

LIVE LOAD $\quad 500 \times 33 \quad\quad = 16,500$

$\quad\quad\quad\quad\quad\quad\quad\quad\quad 87,795 \text{ lbs/ft}$

Bearing Pressure = $\dfrac{87,795 \text{ lb/ft}}{12 \text{ ft}}$ = 7300 lb/ft^2

SINGSTAD, HURKA & ASSOCIATES, P.C.
Consulting Engineers

Frank P Hurka, P.E.
Anthony S. Craven, P.E.
Joseph A. Defino, R.A.
Surendra Sanwasi, P.E.
Simon Zidrinsky, P.E.

August 7, 1984

Brooklyn Historic Railway Association
599 East 7th Street
Brooklyn, New York 11218

Attention: Mr. Robert Diamond, President

RE: The Atlantic Avenue Tunnel

Gentlemen:

On Thursday, July 19, 1984, at your request, the undersigned and Mr. J. A. Defino (Chief Architect, SHA) participated with members of the Brooklyn Historic Railway Association (BHRA) in a preliminary survey to determine the existing condition of "The Atlantic Avenue Tunnel" which was constructed approximately one hundred and forty years ago.

The group entered the abandoned tunnel through an access manhole (at the corner of Atlantic Avenue and Court Street) and proceeded to conduct a visual and general survey of the tunnel; from the existing east bulkhead (west of Court Street) to the west bulkhead (east of Columbia Street), a distance of approximately 1500 feet.

The tunnel cross section is an horseshoe shaped structure approximately 21 feet wide at the base and 17 feet high at the centerline. It is comprised of granite block vertical walls to the springline and then brick masonry for the arched portion of the horseshoe.

An instrumental survey and detailed inspection of the tunnel were not possible due to the limited time available to inspect the site. However, sufficient information was obtained from visual observations to permit the following conclusions:

APPENDIX "A"

SINGSTAD, HURKA & ASSOCIATES, P.C.

Attention: Mr. Robert Diamond August 7, 1984

1. For a tunnel structure which is approximately 140 years of age, the observed portion of the masonry tunnel (1500 feet long) is remarkably free of deterioration and appears to be structurally sound.

2. There is no presence of any inflow of ground water at the invert and the tunnel interior is relatively dry and shows no signs of any water infiltration.

3. There is evidence of moisture droplets on the masonry arch and the fluorescent light fixtures in certain portions of the tunnel. This is due to atmospheric condensation created by the tunnel "opening" in 1980. This can be corrected by installing mechanical ventilation fans to permit fresh air circulation throughout the tunnel.

It should be noted that while the apparent condition of the masonry tunnel and the quality of the workmanship indicates extraordinary technical competence, the state of the art of tunnel design was in its infancy when this tunnel was constructed. An in depth design analysis, using field measured data, would be needed for a definitive determination of any rehabilitation need.

In conclusion, insofar as a preliminary, visual survey can determine, "The Atlantic Avenue Tunnel" is presently structurally sound and quite safe for authorized personnel to enter thereon to conduct further detailed investigations.

Singstad, Hurka & Associates is quite pleased to be part of the restoration program of "The Atlantic Avenue Tunnel" and is ready to assist BHRA in whatever way it can to fulfill the goal of revitalizing Atlantic Avenue and its immediate community.

Very truly yours,

SINGSTAD, HURKA & ASSOCIATES, P.C.

Anthony S. Caserta
Executive Vice President

ASC:fk
cc: Mr. Sebastian Scialabba, P.E.

CITY OF NEW YORK
DEPARTMENT OF GENERAL SERVICES
DIVISION OF MUNICIPAL SUPPLIES
Laboratory
INTRA-DEPARTMENT
MEMORANDUM

TO: Robert Diamond, President
Brooklyn Historic Railroad Association

DATE: Aug 07 ' 85

FROM: A.D. Pacifico, DGS Laboratory Director

SUBJECT: Test Results

925 - 5406

Attached is our laboratory report on samples of cores, brick, and soil taken from the Atlantic Avenue Tunnel.

Professionally, we were glad to have the opportunity to test materials which were used in a construction project dating back to the mid 19th Century.

We hope the data supplied is sufficient for your needs. If additional information is required, please contact me.

We have set aside the test samples and await your instructions, as to their disposition.

Attachment: Lab Report B6 - 0886
ADP: ejw

Sheet #1 of 13 sheets

A QUALITATIVE REVIEW OF CORES RECEIVED FROM

THE ATLANTIC AVENUE TUNNEL IN BROOKLYN

THE FOLLOWING IS A BRIEF DESCRIPTION OF CORES RECEIVED BY THIS LABORATORY FOR EVALUATION. THE CORES WERE TAKEN FROM VARIOUS LOCATIONS IN THE TUNNEL AT ATLANTIC AVENUE BETWEEN HICKS AND COURT STREETS, BROOKLYN.

/CORE #1/
Core #1 is composed of a brick section and a stone section. In the brick section, there are 5 layers of brick with mortar joints, which measure approximately 20 inches. The mortar bond still exixts between each individual brick. Two composite brick samples were taken from this core for test. The results are attached. The stone section has pieces of random lengths. There is one piece, 15 inches, and another 12 inches. The others are fragments of varying sizes from 3 inches to small pebbles. This core was taken from the side of the Tunnel.

/CORE #2/
Core #2 has a brick section only. There are 5 layers of brick with mortar joints. The total length is approximately 20.5 inches. The individual bricks are still bonded together with mortar. One composite brick sample was tested from this section, and one individual brick sample. The test results are attahced. This core was taken from the ceiling of the Tunnel.

/CORE #3/
Core #3 is of brick section and stone section. There are 5 layers of brick with mortar joints and these measure approximately 20 inches. The individual bricks are bonded by mortar joints. In the stone section, there is one piece, 13.5 inches, another 12 inches, and fragments and pebbles of varying sizes. No test sample was taken from this core. The core was obtained from the side of the Tunnel.

/CORE #4/
Core #4 is of brick section only. There are 5 layers of brick with mortar joints measuring approximately 20 inches. The mortar bond still exists between the individual bricks. No test sample was taken from this core. The core was obtained from the ceiling of the Tunnel.

QUALITATIVE REVIEW OF CORES RECEIVED FROM ATLANTIC AVE. TUNNEL - BKN

RE: CORES .... continued ...

/CORE #5/
Core #5 is of brick section and stone section. Brick section measures approximately 20 inches. There are 5 layers of brick with mortar joints. The bond between individual bricks still exists. The stone section has various sizes of stone held together with mortar and measures approximately 22 inches. One composite brick sample and one individual brick were tested from this core. The results are attached. This core was taken from the side of the Tunnel.

/CORE #6/
Core #6 is of brick section only. There are 5 layers of brick bonded by mortar. Individual bricks are also bonded by mortar. Two composite brick samples were tested from this core. This core was taken from the ceiling of the Tunnel.

/CORE #7/
Core #7 is comprised of brick section and stone section. The brick section has 5 layers of brick with mortar joints and measures approximately 20 inches. No test sample was taken from this core. The stone section has an overall length of approximately 23 inches. One piece is 17 inches, and the other pieces are of varying sizes held together by mortar. This core was taken from side of Tunnel.

/CORE #8/
Core #8 is of brick section only. There are 5 layers of brick with mortar joints. Overall measurement is approximately 20 inches. Each individual brick is bonded with mortar. No test sample was taken from this core. The core was obtained from ceiling of Tunnel.

/CORE #DP4/
Core #DP4 is of varying sizes of stone, held together by mortar. The overall length is approximately 62 inches. The stone appears to be Mica Schist, based upon observation of its structure and the amount of mica flakes contained.

Two samples were taken for degradation test by Los Angeles Machine, as per A.S.T.M. C 535.

The results of the tests are attached.

DEPARTMENT OF GENERAL SERVICES
LABORATORY - 480 Canal St., NYC

CITY OF NEW YORK
DEPARTMENT OF GENERAL SERVICES
Bureau of Laboratories

Sheet #3 of 13 Sheets
Date 7-16-85
Lab # B6-0886

Insp. # _____ Lab. # _____
Dept. _____ Sect. _____
Material 4" ⌀ BRICK CORE   Spec. # _____
Brand _____ Contract # _____
Contract ATLANTIC AVE. TUNNEL
Point/Date of Sampling _____
Remarks CORE #1
Sampled by: ROBERT DIAMOND
                    (Signature)

Insp. # _____
Contract # _____

79 - Date Rec'd by Lab.

Specification   Date received in lab 6- 85

TESTS	Found	Requirements
WATER CONTENT, 24 hrs. IN COLD WATER (%)	10.1	
WATER CONTENT, 5 hrs. BOILING, (%)	12.1	
COMPRESSIVE STRENGTH (PSI):		
SAMPLE # 1 COMPOSITE	3,382	
SAMPLE # 1A COMPOSITE	2,626	
SAMPLE #		
SAMPLE #		
SAMPLE #		

CITY OF NEW YORK
DEPARTMENT OF GENERAL SERVICES
LABORATORY - 480 Canal St., NYC

Sheet #4 of 13 Sheets

Date 7-16-85

CITY OF NEW YORK
DEPARTMENT OF GENERAL SERVICES
Bureau of Laboratories

Lab # B6-0886

Insp. # _____ Lab. # _____
Dept. _____ Sect. _____
Material 4" Ø BRICK CORE Spec. # _____
Brand _____ Contract # _____
Contract ATLANTIC AVE. TUNNEL
Point/Date of Sampling _____
Remarks CORE #2
Sampled by: ROBERT DIAMOND
(Signature)

Insp. # _____

Contract # _____

79 - Date Rec'd by Lab.

Specification   Date received in lab 7- 85

TESTS	Found	Requirements
WATER CONTENT, 24 hrs. IN COLD WATER (%)	10.2	
WATER CONTENT, 5 hrs. BOILING, (%)	11.3	
COMPRESSIVE STRENGTH (PSI):		
SAMPLE # 2 COMPOSITE	3,581	
SAMPLE # 2 HALF BRICK	3,076	
SAMPLE #		
SAMPLE #		
SAMPLE #		

CITY OF NEW YORK
DEPARTMENT OF GENERAL SERVICES
LABORATORY - 480 Canal St., NYC

Sheet #5 of 13 sheets

CITY OF NEW YORK
DEPARTMENT OF GENERAL SERVICES
Bureau of Laboratories

Date 7-16-85

Lab # B6-0886

Insp. # _____ Lab. # _____
Dept. _____ Sect. _____
Material 4" ⌀ BRICK CORE   Spec. # _____
Brand _____ Contract # _____
Contract ATLANTIC AVE. TUNNEL
Point/Date of Sampling _____
Remarks CORE #5
Sampled by: ROBERT DIAMOND
(Signature)

Insp. # _____

Contract # _____

79 - Date Rec'd by Lab.

Specification

Date received in lab 7- 85

TESTS	Found	Requirements
WATER CONTENT, 24 hrs. IN COLD WATER (%)	11.3	
WATER CONTENT, 5 hrs. BOILING, (%)	12.5	
COMPRESSIVE STRENGTH (PSI):		
SAMPLE # 5 - COMPOSITE	3,242	
SAMPLE # 5A - HALF BRICK	3,451	
SAMPLE #		
SAMPLE #		
SAMPLE #		

CITY OF NEW YORK
DEPARTMENT OF GENERAL SERVICES
LABORATORY - 480 Canal St., NYC

Sheet #6 of 13 sheets
Date 7-16-85

CITY OF NEW YORK
DEPARTMENT OF GENERAL SERVICES
Bureau of Laboratories

Lab # B6-0886

Insp. # _____ Lab. # _____
Dept. _____ Sect. _____
Material 4" ⌀ BRICK CORE Spec. # _____   Insp. # _____
Brand _____ Contract # _____
Contract ATLANTIC AVE. TUNNEL   Contract # _____
Point/Date of Sampling _____
Remarks CORE #6
Sampled by: Robert Diamond
              (Signature)

79 - Date Rec'd by Lab.

Specification   Date received in lab 7- 85

TESTS	Found	Requirements
WATER CONTENT, 24 hrs. IN COLD WATER (%)	10.3	
WATER CONTENT, 5 hrs. BOILING, (%)	11.9	
COMPRESSIVE STRENGTH (PSI):		
SAMPLE # 6 COMPOSITE	3,183	
SAMPLE # 6A COMPOSITE	2,667	
SAMPLE #		
SAMPLE #		
SAMPLE #		

DEPARTMENT OF GENERAL SERVICES
LABORATORY - 480 Canal St., NYC

CITY OF NEW YORK
DEPARTMENT OF GENERAL SERVICES
Bureau of Laboratories

Date **7-16-85**

I. #_____ Lab. # **B6-0886**
Dept._____ Sect._____
Material **STONE** Spec. #_____
Brand _____
Contract #_____
Contract **ATLANTIC AVE. TUNNEL**
Point/Date of Sampling _____
Remarks **LOS ANGELES MACHINE TEST**
Sampled by: **ROBERT DIAMOND**
(Signature)

Lab. # **B6-0886**

Insp. #_____

Contract #_____

INFORMATION

79 - Date Rec'd by Lab.

Specification

Date received in lab. 7- 85

Tests	Found	Requirements
LOS ANGELES MACHINE TEST		
1- Loss in weight for sample with maximum size of 3 inches	32.45 %	
2- Loss in weight for sample the same size as core was received	7.58 %	

Note:
The Los Angeles Machine Test was run on both the stone sample as received and the stone sample broken into approximately 3" pieces. The higher result for the 3 inch pieces can be explained by the increased surface area and the softer interior of the sample.

(7)

CITY OF NEW YORK
DEPARTMENT OF GENERAL SERVICES
LABORATORY – 480 Canal St., NYC

Sheet #8 of 13 Sheets

Date 7/31/85

CITY OF NEW YORK
DEPARTMENT OF GENERAL SERVICES
Bureau of Laboratories

Insp. # _____ Lab. # _____
Dept. General Services Sect. _____
Material Soil
Brand _____ Spec. # _____
Contract _____ Contract # _____
Point/Date of Sampling Old subway tunnel at Atlantic
Remarks Avenue & Court Street
Sampled by: Joseph
(Signature)

Lab. # B6-0886
Sample 1
Insp. # _____
Contract # _____

INFORMATION

SAMPLE 1 WAS TAKEN FROM CORE LOCATION No. 8

79 - Date Rec'd by Lab. JUL 23 1985

Eleanor C. Eastman

Specification   Date received in lab.

Tests	Found	Requirements
Material	Soil	
Appearance	MEDIUM BROWN IN COLOR CONTAINS STONES OF VARIOUS SIZES	
Natural Water Content, %	3.5	
Sieve Analysis, By weight passing, % Sieve Size:		
1 inch	53.3	
3/8 inch	47.0	
No. 4	42.5	
No. 8	38.0	
No. 40	19.4	
No. 200	4.6	
Plastic Limits, %	NONPLASTIC	
(8)		Paul Kissinger

Sheet #9 of 13 sheets

DEPARTMENT OF GENERAL SERVICES
LABORATORY - 480 Canal St., NYC

Date 7/31/85

CITY OF NEW YORK
DEPARTMENT OF GENERAL SERVICES
Bureau of Laboratories

Lab. # B6-0886
Sample 2

Insp. # _____ Lab. # _____
Dept. General Services Sect. _____
Material Soil _____ Spec. # _____
Brand _____ Contract # _____
Contract _____
Point/Date of Sampling Old subway tunnel at Atlantic
Remarks Avenue & Court Street
Sampled by: Joseph
           (Signature)

Insp. # _____
Contract # _____

INFORMATION

SAMPLE 2 WAS TAKEN FROM CORE LOCATION No. 7
79 - Date Rec'd by Lab. JUL 23 1985

Eleanor C. Eastman

Specification   Date received in lab.

Tests	Found	Requirements
Material	Soil	
Appearance	MEDIUM BROWN IN COLOR CONTAINS STONES OF VARIOUS SIZES	
Natural Water Content, %	4.7	
Sieve Analysis, By weight passing, % Sieve Size:		
1 inch	85.2	
3/8 inch	71.6	
No. 4	61.8	
No. 8	55.1	
No. 40	27.5	
No. 200	6.1	
Plastic Limits, %	NONPLASTIC	

Paul Ricci

GRAIN SIZE ANALYSIS

		PERCENTAGE FINER BY WEIGHT	

BOULDERS	COBBLES	COARSE GRAVEL	MEDIUM GRAVEL	FINE GRAVEL	COARSE SAND	MEDIUM SAND	FINE SAND	SILTS & CLAYS IDENTIFIED BY PLASTICITY
228	76.2	25.4		9.52	2.0	0.59	0.25	0.074 mm
9"	3"	1"		3/8"	No.10	No.30	No.60	No.200 Sieve

Sieve sizes: 3½", 3", 2½", 2", 1½", 1", 3/4", 5/8", 1/2", 3/8, 4, 8, 10, 16, 20, 30, 40, 50, 60, 80, 100, 140, 200

Descriptive terms (right side): trace 0, little 10, some 20, and 35, and 50, some 35, little 20, trace 10, 0

GRAIN SIZE IN MILLIMETERS

LOCATION _____
REMARKS _____

Sample No.	2	Depth
Lab. No.		
Date		

DESCRIPTION _____

LL ____ PL ____ PI ____
SG ____ By ____

Project _____

(11)

DEPARTMENT OF GENERAL SERVICES
LABORATORY - 480 Canal St., NYC

Date 7/31/85

CITY OF NEW YORK
DEPARTMENT OF GENERAL SERVICES
Bureau of Laboratories

Lab. # B6-0886
Sample 3

Insp. # _____ Lab. # _____
Dept. General Services Sect. _____
Material Soil   Spec. # _____
Brand _____ Contract # _____
Contract _____
Point/Date of Sampling Old subway tunnel at Atlantic
Remarks Avenue & Court Street
Sampled by: Joseph
               (Signature)

Insp. # _____

Contract # _____

INFORMATION

SAMPLE 3 WAS TAKEN FROM CORE LOCATION No. 6
79 - Date Rec'd by Lab. JUL 23 1985

Eileen C. Eastman

Specification   Date received in lab.

Tests	Found	Requirements
Material	Soil	
Appearance	MEDIUM BROWN IN COLOR CONTAINS STONES OF VARIOUS SIZES	
Natural Water Content, %	3.2	
Sieve Analysis, By weight passing, % Sieve Size:		
1 inch	76.2	
3/8 inch	69.0	
No. 4	62.6	
No. 8	56.5	
No. 40	25.1	
No. 200	3.6	
Plastic Limits, %	NONPLASTIC	

Paul Kissinger

(12)

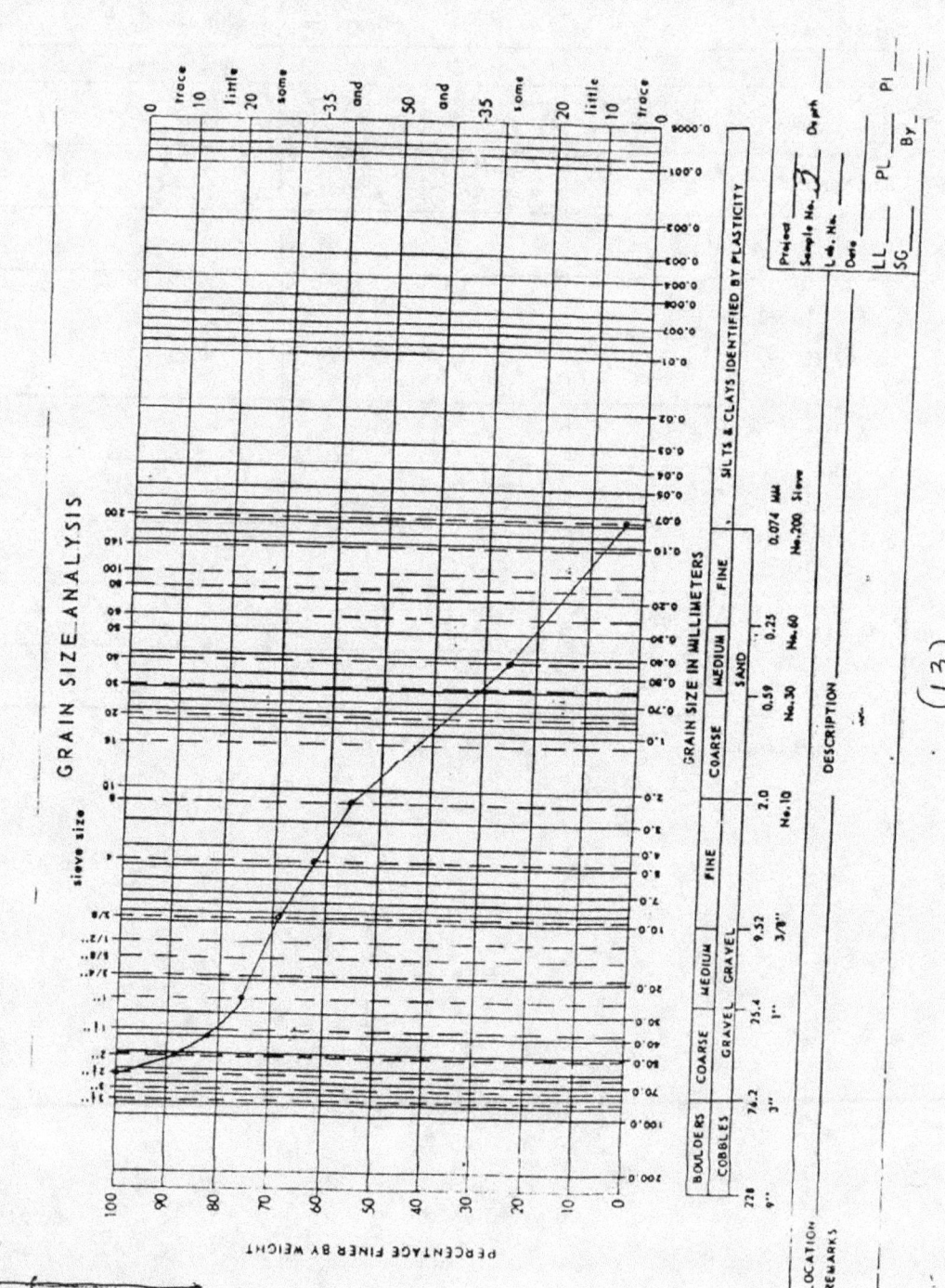

Pre- Existing City Agency Approvals

New York City Transit Authority

370 Jay Street, Brooklyn, New York, 11201 Phone (718) 330-

Members of the Board
Robert R. Kiley
Chairman
Lawrence R. Bailey
Daniel T. Scannell
Vice Chairmen
Stephen Berger
Laura Blackburne
Stanley Brezenoff
Thomas Egan
Herbert J. Libert
John F. McAlevey
Ronay Menschel
Constantine Sidamon-Eristoff
Robert F. Wagner, Jr.
Robert T. Waldbauer
Alfred E. Werner

David L. Gunn
President

January 9, 1986

Mr. Joseph W. Ketas
Director, CEQR
Department of City Planning
Two Lafayette Street, Room 2400
New York, N.Y. 10007

Re: CEQR-85-003K
ATLANTIC AVE. TUNNEL PROPOSAL

Dear Mr. Ketas:

This letter is in response to your December 23, 1985 letter to C. L. Turin enclosing an engineering report submitted by the applicant of the subject project.

We have made a cursory review of the November 27, 1985 report by Singstad, Hurka & Associates submitted with your December 23 letter. We have no comments on this report except to note that on pg. 2 a live load of 500 P.S.F. was used in the design. The Transit Authority uses a live load that varies with depth. The attached Table 3 indicates the loading which, for a cover of 10 feet, is 0.6 K.S.F. The Consultant is experienced in the design of tunnels and can be relied on for a proper recommendation.

Very truly yours,

M. Oberter, P.E.
Division Engineer

320:NK:ss
123085 NK/L/ss
Attachment

Copy To: Mr. Robert Diamond, President
Brooklyn Histroical Railway Assoc.
599 E. 7th Street
Brooklyn, N.Y. 11218

Honorable Jack Lusk
Office of the Mayor
52 Chambers Street (Rm 108)
New York, N.Y. 10007

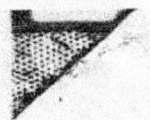

Table 3 gives the uniform live load, together with the corresponding dead load, for various covers of dry earth.

Table 4 gives the equivalent total load per sq. ft. which for various covers, spans and spacings of roof beams produces in the latter the same moment or shear as does the local concentration specified in Subsection (b) above, together with the corresponding dead load, for various covers of dry earth.

TABLE 3

SIDEWALK AND ROADWAY LOAD OVER SUBWAYS IN KIP PER SQ. FT.

Cover in Ft.	Dead Load (Cover Load)	Live Load		Total Load	
		Sidewalk	Roadway	Sidewalk	Roadway
2	0.2	0.6	1.3	0.8	(1.5)
3	0.3	0.6	1.2	0.9	(1.5)
4	0.4	0.6	1.1	1.0	(1.5)
5	0.5	0.6	1.0	1.1	1.5
6	0.6	0.6	0.9	1.2	1.5
7	0.7	0.6	0.8	1.3	1.5
8	0.8	0.6	0.7	1.4	1.5
9	0.9	0.6	0.6	1.5	1.5
10	1.0	0.6	0.6	1.6	1.6
11	1.1	0.6		1.7	
12	1.2	0.6		1.8	
13	1.3	0.6		1.9	
14	1.4	0.6		2.0	
15	1.5	0.5		2.0	
16	1.6	0.4		2.0	
17	1.7	0.3		2.0	
18	1.8	0.2		2.0	
19	1.9	0.1		2.0	
20	2.0	0.0		2.0	

For each additional ft. of cover, increase total load by 0.1 kip per sq. ft.
Values in brackets are minimum values and shall be compared to those given in Table 4.
For roofs below water or with depressed ceiling, increase load as specified in Sec. 2.

SD-3-6

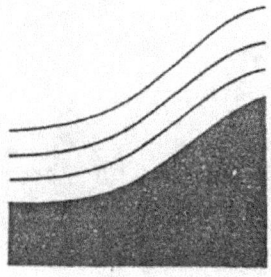

CITY OF NEW YORK
DEPARTMENT OF ENVIRONMENTAL PROTECTION
BUREAU OF WATER SUPPLY

1250 BROADWAY, NEW YORK, N.Y. 10001   (212) 971-6796

JOSEPH T. McGOUGH, JR.
Commissioner

JOSEPH P. CONWAY, P.E., Director
Assistant Commissioner

September 13, 1984

Mr. Morris Tarshis, Director
Bureau of Franchises
Board of Estimate
Municipal Building, Room 1307
New York, New York 10007

           Re: Proposed Entrances to Railroad Tunnel
               on Atlantic Avenue, Borough of Brooklyn.

Dear Sir:

This is in response to your letter dated June 22, 1984 submitting a petition on the above subject. The Bureau was to review the petition to see if there were any objections to maintain and use an abandoned railroad tunnel, including the construction of entrances and facilities under, along and across Atlantic Avenue from east of Court Street to Hicks Street, Borough of Brooklyn.

The Bureau does have objections to these new entrances as shown on the plans. As proposed the petitioner would be required to replace sections of the sixteen (16) inch low pressure main on Court Street; the eight (8) inch low pressure main, the twelve (12) inch high pressure fire main, the twenty (20) inch low pressure main and finally the forty-eight (48) inch trunk main all crossing the proposed entrance at Atlantic Avenue. The cost of these replacements would be approximately one-hundred thousand ($100,000) dollars.

Therefore at Court Street it is suggested that the proposed entrance be placed along the south sidewalk of Atlantic Avenue east of Court Street. This would reduce the water mains to be replaced to sections of a twenty (20) and twelve (12) inch low pressure mains which approximately would cost sixteen thousand ($16,000) dollars.

Mr. Morris Tarshis   9/13/84   Page 2

At Hicks Street the new entrance would require the replacement of two (2) twenty (20) inch high pressure fire mains and one (1) twelve (12) inch low pressure main. The replacement costs of these would be eighteen thousand ($18,000) dollars. The southwest corner site for the proposed entrance does minimize the water main replacement.

As far as the proposed use of the manhole at Court Street this Bureau must object since there is a sixteen (16) inch high pressure water main exposed which would be in danger of being damaged. Therefore this manhole shall have to be removed and the area below backfilled to protect this main.

If the proposal is to be done this Bureau would require a bond to be posted for the replacement water main work. If the proposal is done as the petitioner decribes in his drawing the bond would be one-hundred eighteen thousand dollars ($100,000 for Court Street plus $18,000 for Hicks Street). If the petitioner changes the Court Street location as suggested to the south side of Atlantic Avenue the bond would be thirty-four thousand dollars (Court Street $16,000 plus Hicks Street $18,000).

All water main work shall be done under the inspection of Bureau inspectors after plans have been approved by the Construction Division of the Bureau.

                               Very truly yours,

                               MARTIN E. ENGELHARDT, P.E.
                               Chief, Planning & Programs
                               Bureau of Water Supply

GDeF/lb

bcc: Engelhardt, Dorf, Kushner, Kass, Brooklyn Borough Office
      w/original submission

Schrader

FIRE DEPARTMENT
250 LIVINGSTON STREET   BROOKLYN, N.Y. 11201-5884

BUREAU OF FIRE PREVENTION

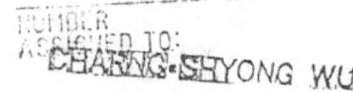

BOARD OF ESTIMATE
BUREAU OF FRANCHISES
RECEIVED

109193 SP 26 84

NUMBER ASSIGNED TO:
CHANG-SHYONG WU

September 25, 1984

Mr. Morris Tarshis
Director Bureau of Franchises
Room 1307
Municipal Building
One Centre Street
New York, N.Y. 10007

    Subject: Bureau of Franchise No. 108062

Dear Mr. Tarshis:

The Fire Department has reviewed the tentative proposal made by the Brooklyn Historic Railroad Association relative to the use of the abandoned railroad tunnel under Atlantic Avenue running from a point West of Boerum Place to a point East of Columbia Street, borough of Brooklyn.

This department will not oppose the project, provided that the safety of the public is paramount. We have discussed with Mr. Robert Diamond, President of the Brooklyn Historic Railroad Association, the requirements essential for public safety. These requirements include the submission of a Fire Protection Plan through the Department of Building from which the Fire Department will receive a copy for review and approval. We have outlined to Mr. Diamond the necessity for automatic sprinklers, a local alarm system, emergency lighting, standpipe, smoke and gas detectors, and forced ventilation system with proper controls and safeguards under fire conditions. We have also indicated the need for additional exit facilities suitable for the use of the public and in accordance with law.

Mr. Diamond has assured this department that he will meet the above requirements and, if possible, exceed them. Mr. Diamond will have a registered architect submit the required building and Fire Protection Plans through the proper channels outlined above. When suitable plans have been submitted and any recommendations for change have been complied with in the interest of public safety, the Fire Department will approve this project.

                                                Very truly yours,

                                                Robert E. Manson
                                                Deputy Assistant Chief
                                                Technology Management
                                                Bureau of Fire Prevention

REM:MJB:mr

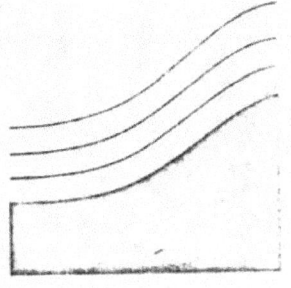

CITY OF NEW YORK
DEPARTMENT OF ENVIRONMENTAL PROTECTION
BUREAU OF SEWERS
40 WORTH STREET, NEW YORK, N.Y. 10013   (212) 566-2104/5

JOSEPH T. McGOUGH, JR
Commissioner

JOHN L. DiMARTINO, P.E., Director
Assistant Commissioner

AUG 23 1984

Morris Tarshis, Director
Bureau of Franchises
Board of Estimate
Room 1307, Municipal Building
New York, NY 10007

Re: Petition by: The Brooklyn
Historic Railway Association,
Brooklyn, K235

Dear Mr. Tarshis:

This is in reply to your letter dated 22 June 1984 which referred to a petition by The Brooklyn Historic Railway Association requesting consent to maintain, and use an abandoned railroad tunnel, including the construction of an entrance and facilities to accommodate the public, under, along and across Atlantic Avenue from east of Court Street to Hicks Street, Borough of Brooklyn.

Please be advised that the Bureau of Sewers has no objection to the petitioner's request. This approval is predicated upon a similar favorable response from the Bureau of Water Supply which is required to comment separately.

Approval by the Department of Environmental Protection is valid only when approvals have been conveyed to your office by both the Bureaus of Water Supply and Sewers.

Very truly yours,

FRANK OLIVERI, P.E.
Acting Director
Bureau of Sewers

ROBERT M. LITKE
COMMISSIONER

CITY OF NEW YORK
DEPARTMENT OF GENERAL SERVICES
DIVISION OF PUBLIC STRUCTURES
MUNICIPAL BUILDING
16TH FLOOR
NEW YORK, N.Y. 10007

GREGORY JOHNSON
DEPUTY COMMISSIONER

September 12, 1984

Re: BROOKLYN HISTORIC RAILWAY ASSOCIATION
Petition dated June 18, 1984 to the
BOARD OF ESTIMATE for consent to
maintain and use an abandoned railroad
tunnel, including the construction of an
entrance and facilities to accommodate
the public under; along and across
Atlantic Avenue from east of Court Street
to Hicks Street
Bureau of Franchises #108062
Borough of Brooklyn

Mr. Morris Tarshis
Director
Bureau of Franchises
1307 Municipal Building
1 Centre Street
New York, N.Y. 10007

Dear Mr. Tarshis:

This is in reply to your June 22, 1984 letter regarding the above matter.

Please be advised that we have no objections to the above consent.

Very truly yours,

Martin Burrell, P.E.
Director
Bureau of Electrical Control

HT/EC:341/ip

BOARD OF ESTIMATE
BUREAU OF FRANCHISES
RECEIVED

108076 SP 14 84

NUMBER
ASSIGNED TO: CHARNG-SHYONG WU

CITY OF NEW YORK
DEPARTMENT OF TRANSPORTATION
OFFICE OF THE COMMISSIONER
40 WORTH STREET • NEW YORK, N.Y. 10013

ANTHONY R. AMERUSO, P.E.
COMMISSIONER

Honorable Morris Tarshis
Director of Franchises
Board of Estimate
Municipal Building
New York, New York 10007

Dear Mr Tarshis:

Regarding the petition to the Board of Estimate from The Brooklyn Historic Railway Association dated June 13, 1984 requesting consent to construct, maintain and use an abandoned railroad tunnel, including the construction of an entrance and facilities to accommadate the public, under, along and across Atlantic Avenue from East of Court Street to Hicks Street, in the Borough of Brooklyn, please be advised that the Department of Transportation has no objection to the petition provided the following conditions are observed:

1) Permits are secured from the Brooklyn Borough Office before starting work.

2) The applicant will restore any existing curb, sidewalk or roadway damaged during construction in accordance with Bureau of Highway Operations Standard Specifications.

3) Within thirty (30) days of completion of construction the petitioner shall submit one set of certified revised "as built drawings" and microfiche card of drawings and related correspondence. Certification to be made by a licensed P.E.

4) The petitioner shall notify utility companies and agencies having existing underground facilities in the proposed construction area for their review and approval.

5) The applicant shall comply with applicable sections of Industrial Code Rule 53 of the State of New York (construction, Excavation and Demolition Operations at or near underground facilities).

6) Inspection Reports must be filed with the Bureau of Highways-Operations Mangement at five year intervals certifying the following:

   A) The structural members were inspected by the engineer within the last six (6) months.

   B) The load carrying capacity is sufficient to support the anticipated loading.

   C) The non-load carrying members have been inspected and are secure.

   D) In addition, a microfiche card of above mentioned "as built drawings" and of related correspomdence shall be submitted by the petitioner.

Very truly yours,

Anthony R. Ameruso, P.E.
Commissioner

Engineer's Cost Estimate - 1985

Steven L. Carroll, P.E.
Consulting Engineer

Telephone 718-356-6245

226 Woodrow Road
Staten Island, New York 10312

February 6, 1985

Mr. Robert Diamond
Brooklyn Historical Railway Association
599 East 7th Street
Brooklyn, New York 11218

Dear Robert,

    Enclosed is a summary of my cost estimate with the appropriate backup calculations and references for all items of construction concerning the Atlantic Avenue Tunnel Project.

    I can be reached at the above address and telephone number if any questions arise or if my professional services are again needed. Good luck on this exciting project.

    Sincerely yours,

Steven L. Carroll, P.E.

SLC:flc
Enc.

cc: S. Scialabba

Mr. Robert Diamond   -2-   February 6, 1985

Summary of Estimate

Mobilization	$ 15,000
Excavation	$ 277,531
Concrete	$ 106,883
Tunnel Work	$ 116,210
Restorations	$ 17,328
Removals	$ 14,050
Misc. Metal Work	$ 38,484
Moisture Protection	$ 6,667
Mechanical	$ 96,791
Electrical	$ 48,000
Utility Protection	$ 128,000
Communications	$ 4,000
Miscellaneous	$ 8,000
Borings and Test Pits	$ 3,446
Bond @ 1%	$ 8,804
Contingency @ 20%	$ 176,078
Adm. and Engineering Fees	$ 125,000
Total Cost of Project =	$1,190,272
SAY	**$1.2 Million**

Note that the entrance cost, which is included above, but consists of a variety of items, is estimated as follows:

    Court St. Entrance, $389,000

    Hicks St. Entrance, $352,000

Atlantic Avenue Tunnel Project
Itemized Cost Estimate

Item	Description		Unit	Quantity	Price	Amount
1.	Mobilization		L.S.	—	—	$15,000
2.	Sheeting		Sq. ft.	7,260	28.51	$206,983
3.	Decking	Timber	B.M.	22,303	1.59	$35,462
		Steel	lbs	9,000	.96	$8,640
4.	Earth Excavation		cu.yd	1,476	7.55	$11,147
5.	Excav. of Exist Tunnel Fill		cu.yd	195	49	$9,575
6.	Stone Ballast		cu.yd	56	27.80	$1,577
7.	Backfill		cu.yd	650	6.38	$4,147
8.	Concrete for Entrances		cu.yd	255	381	$97,155
9.	Concrete for Vents		cu.yd	28	276	$7,728
10.	Steel Reinforcing		lbs	28,454	.65	Included
11.	Sleeves in Concrete		each	100	20	$2,000
12.	Tunnel Flooring		Sq. ft.	10,000	2.00	$20,000
13.	Handrail		L.F.	1,000	16.21	$16,210
14.	Railroad Track		L.F.	1,000	80	$80,000
15.	Saving Excav. Tunn. Fill		Stone	Unknown	5.00	Unknown
16.	Record Stone Block Walls		Sq. ft.	1,268	2.70	$3,424
17.	Entrance Sculpture		L.S.	2	5,000	$10,000
18.	Restoration of Sidewalk		Sq. ft.	840	3.49	$2,932
19.	Restoration of Pavement		Sq. yd.	134	7.25	$972
20.	Concrete Excavation		cu.yd	36	183	$6,588
21.	Pavement Excavation		sq.yd	215	4.90	$1,054
22.	Remove Stone Block Wall		cu.ft.	600	3.07	$1,842
23.	Remove Concrete Bulkhd		cu.ft.	1,512	3.02	$4,566
24.	Entrance Handrails		L.F.	160	34.99	$5,598
25.	Non slip stair treads		L.F.	608	12.97	$7,886
26.	Entrance Hatch Doors		each	2	7,000	$14,000
27.	Vent Stacks @ Entrance		L.F.	56	89.30	$5,000
28.	Steel Ladders @ Vents		lbs	6000	1.00	$6,000
29.	3 ply Waterproofing		Sq. ft.	1,692	1.69	$2,860
30.	4 ply Waterproofing		Sq. ft.	1,692	2.25	$3,807
31.	Fire Line Standpipes		L.F.	1,663	34	$56,542
32.	Exhaust Fans		each	3	4,083	$12,249
33.	Plumbing		L.S.	—	—	$28,000
34.	Electrical a) lighting		L.S.	—	—	$48,000
	b) Emerg Light					
	(included above) c) clock outlets					
	d) Burg Alarm					

Item	Description	Unit	Quantity	Price	Amount
35.	Replacement of Mains	L.S.	—	—	$118,000
36.	Maint. and Supp (utilities)	L.S.	—	—	$10,000
37.	Public Address	L.S.	—	—	$3,000
38.	Telephone	L.S.	—	—	$1,000
39.	Crane to Lower Trolley	L.S.	—	—	$1,000
40.	Refr., Souvenir, & Exhibit Stds	L.S.	—	—	$7,000
41.	Borings (4) + Test Pits (8)	L.S.	—	—	$3,446
42.	Bond @ 1%	L.S.	—	—	$8,804
43.	Contingency (20%)	L.S.	—	—	$176,078
44.	Adm. and Engineering Fees	L.S.	—	—	125,000
			TOTAL	=	$1,190,272

Notes in margin: "utility posted" (item 36), "Misc. | Grnm." (items 37–41)

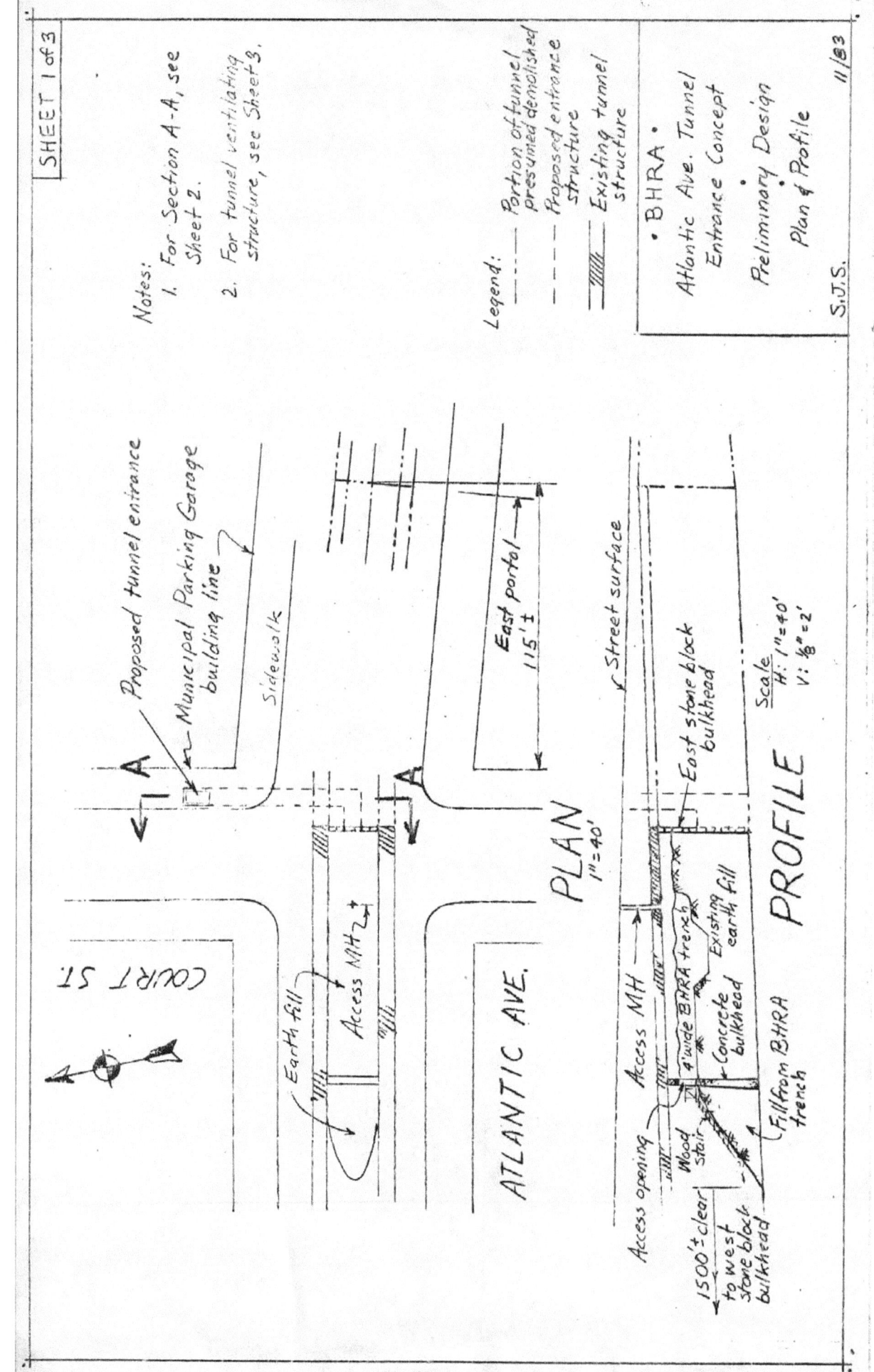

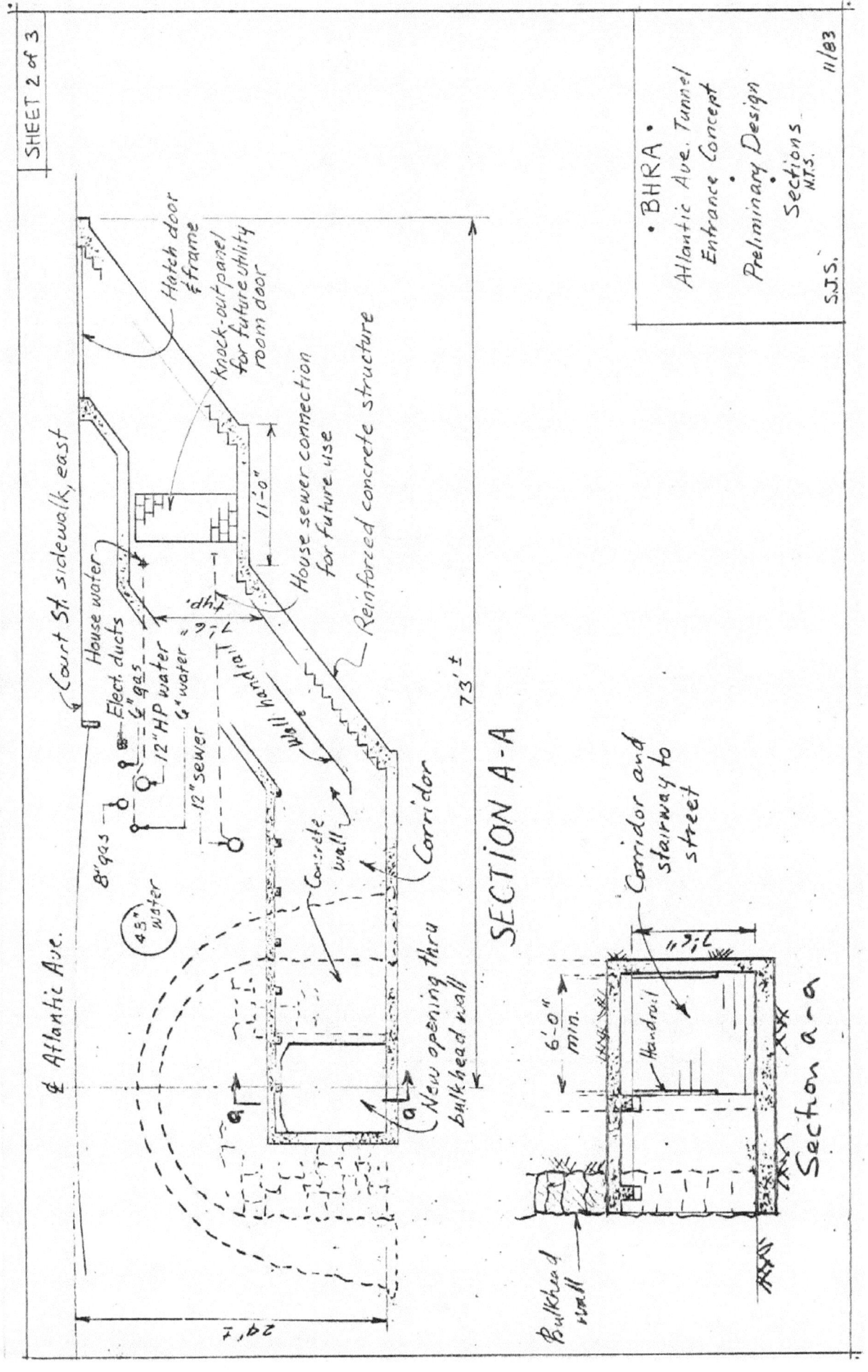

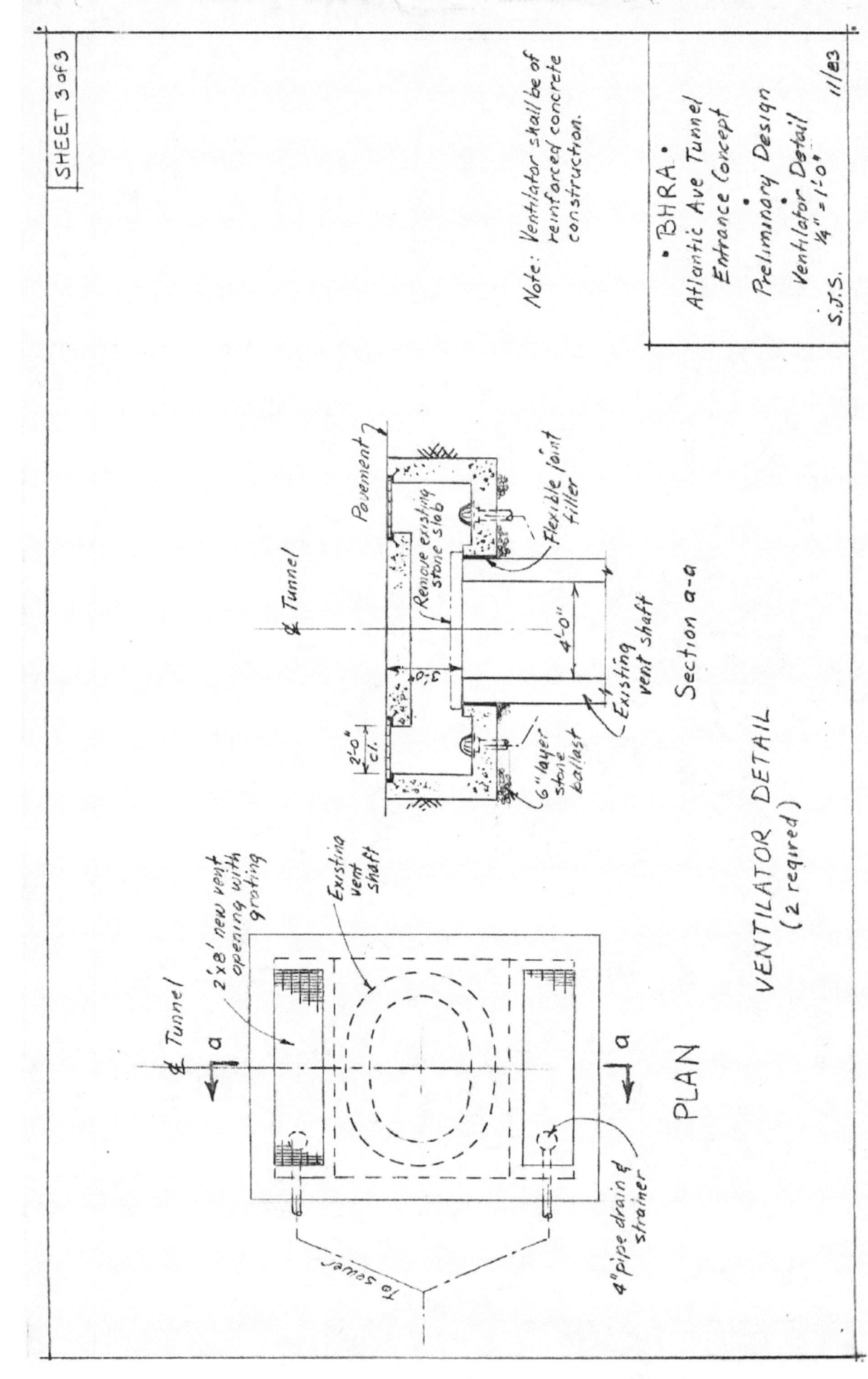

Item	Description	Unit	Quantity	Price	Amount
1.	Mobilization	L.S.	—	—	$15,000
2.	Sheeting	Sq. ft.	7,260	28.51	$206,983
3.	Decking Timber	B.M.	22,303	1.59	$35,462
	Steel	lbs	9,000	.96	$8,640
4.	Earth Excavation	cu yd	1,476	7.55	$11,147
5.	Excav. of Exist Tunnel Fill	cu yd	195	49	$9,575
6.	Stone Ballast	cu yd	56	21.80	$1,577
7.	Backfill	cu yd	650	6.38	$4,147
8.	Concrete for Entrances	cu yd	255	381	$97,155
9.	Concrete for Vents	cu yd	28	276	$7,728
10.	Steel Reinforcing	lbs	28,454	.65	Included
11.	Sleeves in Concrete	each	100	20	$2,000
12.	Tunnel Flooring	Sq. ft.	10,000	2.00	$20,000
13.	Handrail	L.F.	1,000	16.21	$16,210
14.	Railroad Track	L.F.	1,000	80	$80,000
15.	Saving Excav. Tunn. Fill	Stone	Unknown	5.00	unknown
16.	Record Stone Block Walls	Sq. ft.	1,268	2.70	$3,424
17.	Entrance Sculpture	L.S.	2	5,000	$10,000
18.	Restoration of Sidewalk	Sq. ft.	840	3.49	$2,932
19.	Restoration of Pavement	Sq. yd.	134	7.25	$972
20.	Concrete Excavation	cu yd	36	183	$6,588
21.	Pavement Excavation	sq. yd	215	4.90	$1,054
22.	Remove Stone Block Wall	cu. ft.	600	3.07	$1,842
23.	Remove Concrete Bulkhd	cu. ft.	1,512	3.02	$4,566
24.	Entrance Handrails	L.F.	160	34.99	$5,598
25.	Non slip stair treads	L.F.	608	12.97	$7,886
26.	Entrance Hatch Doors	each	2	7,000	$14,000
27.	Vent Stacks @ Entrance	L.F.	56	89.30	$5,000
28.	Steel Ladders @ Vents	lbs	6000	1.00	$6,000
29.	3 ply Waterproofing	Sq. ft.	1,692	1.69	$2,860
30.	4 ply Waterproofing	Sq. ft.	1,692	2.25	$3,807
31.	Fire Line Standpipes	L.F.	1,663	34	$56,542
32.	Exhaust Fans	each	3	4,083	$12,249
33.	Plumbing	L.S.	—	—	$28,000
34.	Electrical a) Lighting	L.S.	—	—	$48,000
	b) Emerg Light				
	(included above) c) Clock outlets				
	d) Emerg. Alarms				

Item	Description	Unit	Quantity	Price	Amount
35.	Replacement of Mains	L.S.	—	—	$118,000
36.	Maint. and Supp. (utilities)	L.S.	—	—	$10,000
37.	Public Address	L.S.	—	—	$3,000
38.	Telephone	L.S.	—	—	$1,000
39.	Crane to Lower Trolley	L.S.	—	—	$1,000
40.	Refr, Souvenir, & Exhibit Stds	L.S.	—	—	$7,000
41.	Borings (4) + Test Pits (8)	L.S.	—	—	$3,446
42.	Bond @ 1%	L.S.	—	—	$8,804
43.	Contingency (20%)	L.S.	—	—	$176,078
44.	Adm. and Engineering Fees	L.S.	—	—	$125,000
			TOTAL	=	$1,190,270

Items 35: utility project
Items 36–38: Comm.
Items 39–44: Misc.

45 | Exterminator | | ? | | ?

COSTS

1982 Means　　　　　　　　　　　　　　　　　　　　　Steven L. Carroll
　　　　　　　　　　　　　　　　　　　　　　　　　　　12/5/83

Concrete:

　　Bms - 2 cuyds
　　Roof - 27 cuyds

　　　　　　　　　　　　　　　　　M fact escal
82　　　29 cuyds × $345/cuyd × 1.15 × 1.2　= $13,807
3-275

82　　　Mat - 37 cuyds × $190/cuyd × 1.15 × 1.2 = $9,701
3-400

82　　　Walls - 46 cuyds × $300/cuyd × 1.15 × 1.2 = $19,044
3-427

　　　　Vent - 28 cuyds × $200/cuyd × 1.15 × 1.2 = $7,728

　　　　　　　　　　　　　　　　　Total = $50,280

Sheeting:

33　　　3080 sq.ft. × $22/sq ft. × 1.08 × 1.2 = $87,817
3-40-450

Excavation:

　　　　　　(+54 cuyds)
29　　　609 cuyds × $1.98/cuyd × 1.08 × 1.2 = $1,563
3-025　　　　　　　　　↓ heavy traffic　　　+$139
　　　hauling -　　(1.3 ×)　　　　　　　　+$269
32　　　609 cuyds × ($2.96/cuyd × 1.08 × 1.2 = $3,037
3-050　　(+54 cuyds)　　　　　　　　　　$4,600 + $408
+140　　　　　　　　　　　　　　　　　　　= $5,008

Back-filling:

27　　　264 cuyds × $4.92/cuyd × 1.08 × 1.2 = $1,683
3-200

Stone Ballast:

P.31
2.3-110
& 320

$26 \text{ cuyds} \times \$14.30/\text{cuyd} \times 1.08 \times 1.2 = \$723$ (1.5×) compaction

Pavement Removal:

P.26
2.1-170
-P.25

$95 \text{ sqyds} \times \$1.26/\text{sqyd} \times 1.08 \times 1.2 = \$465$ (3×) — congested site & small amt.

Concrete Removal:
(Street)

P.25
2.1-030
& -039

$16 \text{ cuyds} \times 3 \times \$47/\text{cuyd} \times 1.08 \times 1.2 = \$2,924$

Roadway Decking:
(12"×12" Timbers)

P.153
6.1-140

$810 \text{ cu ft} \times \dfrac{LBF}{.0825} \times \$1.18/BF \times 1.12 \times 1.2 = \$15,571$

(12 WF 25#)

P.131
:.1-151

$4,250 \text{ lbs}/2000 \text{ lbs} \times \$1,475/\text{ton} \times 1.08 \times 1.2 = \$4,062$

$\qquad\qquad\qquad\qquad\qquad\qquad\qquad\qquad\qquad \$19,633$

Removal of Concrete Bulkhead:

P.25
2.1-010

$(756 \text{ cu ft})/27 \times \$63/\text{cuyd} \times 1.08 \times 1.2 = \$2,286$

Removal of Stone Wall:

P.26
2.1-140

$300 \text{ cu ft} \times \$2.37/\text{cu ft} \times 1.08 \times 1.2 = \$921$

Tunnel Flooring: (OMIT Calculations Below)
(4"×8")   — See Pages 10 & 11 —

P.158
2-020

$760 \text{ L.F.} \times \$4.07/\text{ft} \times 1.12 \times 1.2 = \$4,157$

(plywood) 5/8"

156-010

$1200 \text{ sq ft} \times \$.71/\text{sq ft} \times 1.12 \times 1.2 = \$1,145$

$\}\ \$5,302$

Reinforcement:

P.78 Mat − 4,534 lbs / 2000 lbs × $915/ton × 1.18 × 1.2 = $2,937
3.2-060

P.78 Roof − 2,748 lbs / 2000 lbs × $840/ton × 1.18 × 1.2 = $1,634
3.2-040

P.77 Beams − 2,975 lbs / 2000 lbs × $1,100/ton × 1.18 × 1.2 = $2,317
3.2-010

P.78 Walls − 3,142 lbs / 2000 lbs × $835/ton × 1.18 × 1.2 = $1,858
3.2-010

Total = $8,746

Restoration of Sidewalk:

(6")
P.49 Concrete − 420 sqft × $2.37/sqft × 1.08 × 1.2 = $1,290
2.6-040

P.49 (4")
2.6-045 Stone − 420 sqft × $.32/sqft × 1.08 × 1.2 = $174

$1,464

Restoration of Pavement:

P.47 67 sqyds × $5.60/sqyd × 1.08 × 1.2 = $486
2.6-010

Handrail:

P.138 80 ft. × $27/ft. × 1.08 × 1.2 = $2,799
5.4-240

Stair Treads:

(4" wide)
P.86 38 tr × 8 ft/tr × $9.40/ft. × 1.15 × 1.2 = $3,944
3.3-002

Recondition Stone Block Wall:

P.372
18.1-055

634 sq ft. × $2.25/sq ft. × 1.0 × 1.2 = $1,712

Excavating Tunnel fill:

Hand excavate —

P.30
2.3-140

195 cu yds × $30/cu yd × 1.08 × 1.2 = $7,582

Backhoe Rental —

*150
P.8
1.5-020

1 Day/36 cu yds × 195 cu yds × $5,650/mo × 1/30 days

out put of hand Excavators above
(Use 6 Laborers) ∴ daily output = 36 cu yds

= $1,020

Inside over

Labor (oper Eng + Oiler) —

1 day/36 cu yds × 195 cu yds × $25.65/hr × 7 hrs/1 day

= $973

Total = $9,575

---

3 & 4 ply water proofing:

3 ply

P.172
7.1-070

736 sq ft. × $1.42/sq ft. × .99 × 1.2 = $1,242

4 ply

736 sq ft × $1.89/sq ft. × .99 × 1.2 = $1,653

$2,895

Borings:

P.24
2.1-080

2 × 50' × $9.45/ft. × 1.08 × 1.2 = $1,225

Test Pits:

P.26
2.1-050
+010

Avg 4 × $\frac{(3'\times 3'\times 6')}{27}$ × $48/cu yd × 1.08 × 1.2 = $498

Electrical Work:

$$\$244,381 \times \frac{\$200,000 \text{ (Elect. work on similar proj.)}}{3,000,000 \text{ (Gen Const. Cost on similar proj.)}} \cong \$16,000$$
(Gen Const. Cost) This project

Plumbing Work:

$$\$244,381 \times \frac{\$121,000 \text{ (")}}{\$3,000,000 \text{ (")}} \cong \$9,000$$

USE $25,000

Public Address:

P. 355
16.8-030 Microphone — 1 × $65 → $65
 " -040 Speakers — 3 × $72 → $216
 " -100 Monitor Panel — 1 × $185 → $185
 " -140 Volume Control — 1 × $51 → $51
 " -160 Amplifier — 1 × $925 → $925
 " -180 Cabinets — 1 × $525 → $525

$$\$1,967 \times 1.06 \times 1.2 = \$2,502$$

USE $3,000

(For total see Summary Sheet)

1/10/84   6
Steven L. Carroll

1982 Means
Reference for prices are in the "costs" section of these calculations
2nd Ensurence

Concrete — $42,552

Sheeting — $87,817

Excavation — $4,600

Backfilling — $1,683

Stone Ballast — $723

Pavement removal — $553

Concrete removal — $2,924

Roadway Decking — $19,633

Removal of concrete bulkhead — $2,286

Removal of Stone wall — $921

Restoration of Sidewalk — $1,464

Restoration of pavement — $486

Handrail — $2,799

Stair treads — $3,944

3 & 4 ply waterproofing — $2,895

Borings — $1,225

Test pits — $498

Electrical = $48,000

Plumbing = $28,000

___

Additional ~~30~~ 22 ft. of corridor —

<u>Concrete</u>

Mat — $\dfrac{1' \times 22' \times 11'}{27} \times \$190/\text{cuyd} \times 1.15 \times 1.2 = \underline{\$2,350}$

Bms — $4 \text{ Bms} \times \dfrac{9" \times 7.5' \times 8"}{12 \times 27 \times 12} \times \$345/\text{cuyd} \times 1.15 \times 1.2 = \underline{\$265}$

Roof — $\dfrac{1' \times 10' \times 22'}{27} \times \$345/\text{cuyd} \times 1.15 \times 1.2 = \underline{\$3,879}$

Walls — $1' \times \dfrac{8.25' \times 22'}{27} \times 2 \times \$300/\text{cuyd} \times 1.15 \times 1.2 = \underline{\$5566}$

<u>Sheeting</u>

32 ft. drive for 25 ft. excavation —

$25 \text{ft.} \times 22 \text{ft.} \times 2 \text{ sides} \times \$22/\text{sqft.} \times 1.08 \times 1.2 = \underline{\$31,363}$

<u>Excavation</u>

$\dfrac{10 \text{ft.} \times 25 \text{ft.} \times 22 \text{ft.}}{27} \times (\$1.98 + 1.3 \times \$2.96)/\text{cuyd} \times 1.2 \times 1.08 = \underline{\$1539}$

Backfill

$$\frac{15\text{ ft.} \times 10\text{ ft.} \times 22\text{ ft.}}{27} \times \$4.92/\text{cuyd} \times 1.08 \times 1.2 = \underline{\$779}$$

Stone Ballast:

$$\frac{6" \times 10' \times 22'}{12 \times 27} \times \$14.30/\text{cuyd} \times 1.5 \times 1.08 \times 1.2 = \underline{\$113}$$

Pavement Removal

$$\frac{10' \times 22'}{9} \times \$1.26/\text{cuyd} \times 3 \times 1.08 \times 1.2 = \underline{\$120}$$

Concrete Removal

$$\frac{6" \times 10' \times 22'}{12 \times 27} \times \$47/\text{cuyd} \times 1.08 \times 1.2 = \underline{\$248}$$

Roadway Decking

Timber 2×12
$$\frac{10' \times 22' \times 1\text{ ft.}}{.0825 \text{ cuft./B.F.}} \times \$1.18/\text{B.F.} \times 1.12 \times 1.2 = \underline{\$4,229}$$

2WF25
$$\frac{2\text{ BMS} \times 10' \times 25\text{ plf}}{2000\text{ lbs}} \times \$1,475/\text{ton} \times 1.08 \times 1.2 = \underline{\$478}$$

3 & 4 ply Waterproofing

10 ft. × 22 ft. × $1.42/sq.ft. × .99 × 1.2 = $371

(Total for 22 ft. of Additional Corridor) —

= $51,300

(Total for Duplicate or 2nd Entrance) —

= $224,003

TOT (Entire 2nd Entrance) = $275,303

Tunnel Flooring

<u>Delete</u> the item "Tunnel Flooring" in the previous calculations in the "Costs" section — ($5,302) + 20% (cont.) = ($6,362)

<u>add the following</u>:

[Diagram: cross-section showing 3/4" plywood on top, with 2"×6" (typ) supports spaced 3'-4" apart]

<u>Sears 1982</u> Platform 10 ft. wide × 1000 ft. long

$2" \times 6"$'s — 3'-4" o.c. or 4 in 10 ft.

4 × 1,000 ft. = 4,000 L.F. or B.F.

<u>3/4" plywood</u>

10 ft. × 1,000 ft. = 10,000 sq. ft.

                                        (NY fact) (escal) to 1984

P. 152 $.90/B.ft. × 4,000 B.ft. × 1.12 × 1.2 = $4,838
6.1-410

P. 155 $.82/sq.ft. × 10,000 sq.ft. × 1.12 × 1.2 = $11,021
6.1-030

                     <u>Total for tunnel flooring = $15,859</u>
                     without cost of creosote
                           or fire retarder

Wood treatments:

(material only) —

Fire Retardant —

P.154   2"×6", $\$.150$/B.F. × 4,000 B.F. × 1.12 × 1.2 = $\$806$
6.1-040

P.154   3/4" ply, $\$.18$/sq.ft. × 10,000 sq.ft. × 1.12 × 1.2 = $\$2,419$
6.1-010

$$\text{Fire Ret. TOT} = \$3,225$$

Creosote —

P.154   2"×6", $\$.15$/B.F. × 4,000 B.F. × 1.12 × 1.2 = $\$806$
6.1-030

$$\text{TOT. Treatment} = \$4,031$$

$$\text{TOT for Treated Tunnel Floor} = \$19,890$$

Metal Railing (mild steel, economy)

P.138 $\$12.55$/ft. × 1,000 ft. × 1.076 × 1.2 = $\$16,205$
5.4-191

N.Y.C. Building Code
Standards for Natural Ventilation
(Article 12, Subarticle 1205.0)

Floor Area:

Tunnel — 21 ft. × 1,667 ft. = 35,007 sq. ft.
Entrances — (6 ft. × 76.75 ft. + 8 ft. × 7.5 ft.) × 2
  + 7.5 ft. × 22 ft. = 1,206 sq. ft.

  Total Floor Area = 36,213 sq. ft.

paragraph
C-26-
1205.5

Min Ventilation Area:

A_{min} = .05 × 36,213 sq. ft. = 1,811 sq. ft.

Para C26-1205.6 — $\frac{1}{2}$ of above area is permissable

  if mechanical means of ventilation

  is used:

∴ Min Ventilation Area = $\frac{1811 \text{ sq. ft.}}{2}$
  = 906 sq. ft.

Area furnished:

Tunnel Vents = $3\pi ab$ (ellipse) = 3 Vents × π × 4 ft. × 6 ft.
  = 226 sq. ft.
2 Entrance doors = 2 × 12.5 ft. × 8 ft. = 200 sq. ft.

Para C26-1205.6 — Min Vent opening = 3 sq. ft.

∴ Install 4 Vents - 2 ft. ∅ each
  - Ductile Iron -

$$A = \frac{\pi D^2}{4} = \frac{\pi (2ft.)^2}{4} = 3.14 \text{ sq.ft.} > 3 \text{ sq.ft.}$$
O.K.

$A(\text{Vents}) = 4 \times 3.14 \text{ sq.ft.} = 12.6 \text{ sq.ft.}$

Total Vent. Area Furnished = 438.6 sq.ft. > 906 ?

N.G.

∴ Use Mechanical means as supplemental Ventilation

Standards for Mechanical Ventilation

Article 12, Subarticle 1206.0

Vent. Index:
Para C26-1206.2   Max Occup. = $\frac{A_{min}}{V.I.} \times 200$

Max Occup = $\frac{906 \text{ sq.ft.} \times 200}{301}$ = __601 persons__

Para C26-1206.3, table 12-2

Req'd Supply = 2 cfm/sq.ft.

$Q_{in} = 2 \text{ cfm/sq.ft.} \times 36,212 \text{ sq.ft.} = 72,424 \text{ cfm}$

Req'd Exhaust = 1.5 cfm/sq ft × 36,212 sq. ft.
$$= 54,318 \text{ cfm}$$

$Q_{out} = 54,318$ cfm req'd

furnished:

3 fans @ Q_{out} = 21,600 cfm each

Q_{out} furnished = 3 × 21,600 cfm = 64,800 cfm
> 54,318 cfm req'd

O.K.

Exhaust fans

21,600 CFM, 40 sq in damper

P. 327 3 fans × $3,250/fan × 1.047 × 1.2 = $12,250
5.7-730

Railroad Track

100 lb rail
6" × 8" × 8'-6" wooden ties
6" of stone ballast

P. 65 $58.05/L.F. × 1000 ft. × 1.077 × 1.2 = $75,024
27

2nd distribution, add ⟶ $5,000

TOT R.R. Track = $80,000

Utility connections

water and electric hook up - $1,000 each

TOT = $2,000

Vent Stacks

?.44   4 Vents × 14 ft./Vent × $59/ft. × 1.077 × 1.2 = $4,270
2.5-149

plates, bolts, etc - say $5,000

Sprinkler Requirements
(Art. 3)

occupancy classification — F-3

as per subarticle 307.0, para C26-307.4

construction group — I-A or I-B or at least I-C

as per sub art - 314.0, paras C 26-314.2, .3, .4

__No Sprinkler Req'd__ as per

article 4, table 4-1, Page 4-4

Fire Standpipe Requirements

(Article 17)

__No. of Standpipes required__ — subarticle 1702.0
para C 26-1702.4

125 ft. of hose (max.)
20 ft. of water stream (min.)

Total radius = 145 ft.

distance in both directions from standpipe
= 290 ft. between standpipes (MIN)

2 entrances + tunnel = 1,751 ft.

__No. of Std pipes Req'd__ = $\dfrac{1{,}751 \text{ ft.}}{290 \text{ ft./st pipe}}$ = 6

Size of Standpipe Req'd
(article 17)

Subarticle 1702.0, para C26-1702.7, and table 17-1

hieght < 150 ft. ∴ <u>Min ⌀ = 4"</u>

Reference Standard RS-17
table RS 17-1-1

hieght < 115 ft.

∴ Use - 2½" hose & outlet valves, class A
      pipe - Schedule 40 mild steel
      fittings - steel (350 psi)
      Valves - check or gate valves (150 psi)

Design

6 class A hoses, racks, etc
1,645 ft. of schedule 40 steel pipe (4"⌀)
10 steel tees (4"⌀)
10 steel elbows (4"⌀)
18 ft. of branch pipe 4"⌀ sched 40
7 - 4" steel valves

pour 1ft. × 1ft. cradle (conc.) 1,600 long

Fire Standpipe

P. 293 15.1-191	$(1,645 + 18)\,ft. \times \$14.15/ft. \times 1.047 \times 1.2 = \$29,565$
P. 293 15.1-480	$10 \times \$50/tee \times 1.047 \times 1.2 = \$628$
P. 293 15.1-480	$10 \times \$35/elbow \times 1.047 \times 1.2 = \$440$
P. 299 15.1-195	$7 \times \$840/valve \times 1.047 \times 1.2 = \$7,388$
P. 307 15.4-410	$6 \times \$180/hose\ cabinet \times 1.047 \times 1.2 = \$1,357$
P. 308 15.4-238	$125\,ft.\ of\ Hose \times 2 \times 6 \times \$1.80/ft.\\ \times 1.047 \times 1.2 = \$3,392$
P. 82 3.3-390	$1600\,ft.\ (conc.\ cradle) \times \dfrac{1' \times 1'}{27\,ft.} \times \$170/cu\,yd\\ \times 1.152 \times 1.2 = \$13,926$

$$\text{Total} = \$56,696$$

Quantities

Steven L. Carroll
12/5/83

Concrete: **TOTAL – 126 cu yds**
(inch vent shaft)

Mat
1st Stair –
$1' \times 21' \times 11' = 231$
1st Landing –
$1' \times 10' \times 11' = 110$
2nd Stair –
$1' \times 18.6' \times 11' = 205$
2nd Landing –
$1' \times 32' \times 11' = 352$
Entrance –
$1' \times 8' \times 11' = \underline{88}$
$986/27 \longrightarrow$ 37 cu yds

Roof
1st Stair –
$1' \times 6' \times 10' = 60$
1st Landing –
$1' \times 10' \times 10' = 100$
2nd Stair –
$1' \times 18.6' \times 10' = 186$
2nd Landing –
$1' \times 28' \times 10' = 280$
Entrance –
$1' \times 8' \times 10' = 80$
Drop: $2 \times 10' \times 1' = 20$ $726/27 \longrightarrow$ 27 cu yds

Beams – 8×18", 12" incl above, use 6"×8"
19 bms × $\frac{6"\times 8"}{144"} \times 8' = \underline{51/27} \longrightarrow$ 2 cu yds

Walls
1st Stair –
$(14\frac{1}{2}' \times 12')/2 + 2' \times 12' = 111$
1st Landing –
$10' \times 8' \times 1' = 80$
2nd Stair –
$15 \times (8' + 8')/2 \times 1' = 120$
2nd Landing –
$8' \times 22' \times 1' = 176$
Entrance –   (seperate) $551 \times 2/27 = 41$ cu yds
$8' \times 8' \times 1' = 64, 8 \times 10 \times 1 = 80, 8 \times 8 \times 1 = 64$ + 5 cu yds

Reinforcement: TOTAL – 13,400 lbs

Mat –

For
(splices)   top & bott
Long – 89.6' × 1.1 × 11 bars × (#5) 1.043 plf × 2 = 2,262 lbs
Short – 11' × " × 90 bars × (#5) 1.043 " × 2 = 2,272 lbs
                                                                    4,534 lbs

Roof –

(splices)   Top & bott
Long – 73' × 1.1 × 10 bars × (#4) .668 plf × 2 = 1,073 lbs
Short – 10' × " × 73 bars × (#5) 1.043 plf × 2 = 1,675 lbs
                                                                    2,748 lbs

Beams –

19 beams × 8' × 4 bars × (#9) 3.4 plf = 2,067 lbs
10" × 52"/12 × 19 beams × (#4) .668 plf = 908 lbs
8/12   × 1.1   2,975 lbs

Walls –

1st Stair –
Horz – 16'/2 × 12 bars × (#5) 1.043 plf × 1.1 = 110 lbs
Vert – 13/2 × 16 bars × (#5) 1.043 plf × 1.1 = 119 lbs
1st Landing –
Horz – 10' × 8 bars × (#5) 1.043 plf × 1.1 = 92 lbs
Vert – 9' × 10 bars × (#5) 1.043 plf × 1.1 = 103 lbs
2nd Stair –
Horz – 18.6' × 8 bars × (#5) 1.043 plf × 1.1 = 171 lbs
Vert – 9' × 19 bars × (#5) 1.043 plf × 1.1 = 196 lbs
2nd Landing –
Horz – 32' × 8 bars × (#5) 1.043 plf × 1.1 = 294 lbs
Vert – 9' × 32 bars × (#5) 1.043 plf × 1.1 = 330 lbs
Entrance –
Horz – 8' × 8 bars × (#5) 1.043 plf × 1.1 = 73 lbs
Vert – 9' × 8 bars × (#5) 1.043 plf × 1.1 = 83 lbs
                                                    1571 × 2 = 3,142 lbs

Sheeting: TOTAL – 3080 sq.ft.

1st Stair –
$(15\frac{1}{2}' \times 14')/2 + 2' \times 14' = 137$ sq.ft.

1st Landing –
$8' \times 14' = 112$ sq.ft.

2nd Stair –
$(14' + 26')/2 \times 15' = 225$ sq.ft.

2nd Landing –
$22' \times 26' = 572$ sq.ft.

Entrance –
$8' \times 26' = 208$ sq.ft. $\quad\overline{1,254 \text{ sq ft} \times 2 = 2,508 \text{ sq.ft.}}$
$10' \times 26' = 260$ ⎫ separate
$10' \times 26' = 260$ ⎭ $\longrightarrow\ + 572$ sq.ft.
$2 \times 26' = 52$

Roadway Decking:

$\dfrac{12" \times 12"}{144 \text{ ft}^2} \times 10' \times 81 \text{ timbers} = 810$ cu.ft.

$\approx 170' \times \overset{12wf}{25 \text{ plf sect}} = 4,250$ lbs

Concrete for Vent Shaft:

$52' \times 4.3' \times 1' = 224$ cu ft
$9' \times 9' \times 1' = 81$ cu ft
$2 \times 3\frac{1}{2}' \times 9' \times 1' = 63$ cu ft
$4/12 \times (16' + 20') \times 1' = \underline{12 \text{ cu ft}}$
$\qquad\qquad 380/27 = 14$ cu yds $\times$ 2 Vents = 28 cu yds

Excavating Existing Tunnel Fill: Total = 195 cu yds

Between bulkheads –
$\frac{1}{2}(20' + 20') \times 100' \times 21' = 4200$ cu ft

After 2nd bulkhead –
$(15' + 8')/2 \times 20' + (8 \times 21)/2 \times 20/2 = \underline{1070 \text{ cu ft}}$
$\qquad\qquad\qquad\qquad\qquad\qquad\qquad \dfrac{5270 \text{ cu ft}}{27} = 195$ cu yds

3 & 4 ply Water proofing:

$10' \times (6' + 10' + 18.6' + 28' + 8' + 3') = 736$ sq.ft.

Restoration of Sidewalk:

$12' \times 35' = 420$ sq ft

Restoration of Pavement:

$50' \times 12' = 600$ sq.ft. $\times 1/9$ sq.ft. $= 67$ sq yds

Handrail:
(3 ft high)

2 sides $\times (18.6' + 21') \approx 80$ ft.

Non Slip Stair Treads:
(8' long)

$31'/(10''/12) \approx 38$ treads

Removal of concret Bulkhead:

$(3'+1')/2 \times 18' \times 21' = 756$ cu ft

Entrance opening in stone wall:

$10' \times 10' \times 3' = 300$ cu ft.

Recondition Stone Block walls:

$[9' \times 21' + (\pi \times 21^2)/8] \times 2$ walls $= 724$ sq ft $-$ (opening) 90 sq ft
$= 634$ sq.ft.

Entrance sculpture:

$[((5' + 3½')/2) \times 9' \times 2 + \pi(20^2 - 8^2)/8] \times 2$ walls $= 237$

OMIT ABOVE Calculations

Tunnel Flooring: & Below

$31 \times (4'' \times 8'')/144 \times 18.7' + 2 \times (4'' \times 8'')/144 \times 60' = $ (use 760 L.F.)

Plywood 5/8" $- 60' \times 20' = 1200$ sq ft

Backfill: Total = 264 cu yds

1st Landing –
10' × 3' × 12' = 360 cu ft

2nd Stair –
(3' + 15')/2 × 12' = 108 cu ft

2nd Landing –
15' × 29' × 12' = 5220 cu ft

Entrance –
15' × 8' × 12' = 1440 cu ft

7128/27 = 264 cu yds

Stone Ballast: Total = 26 cu yds

1st Stair –
6"/12 × 21' × 11' = 116 cu ft

1st Landing –
6"/12 × 10' × 11' = 55 cu ft

2nd Stair –
6"/12 × 18.6' × 11' = 102 cu ft

2nd Landing –
6"/12 × 32' × 11' = 176 cu ft

Entrance –
6"/12 × 9' × 11' = 50 cu ft

499/27 ≈ 20 cu yds + 6 cu yds (2 Vents)
= 26 cu yds

Excavation: Total = 609 cu yds

1,566 sq ft × 10.5 ft = 16,443 cu ft / 27 = 609 cu yds

2 Vents – (11 × 15 × 4.33')/27 = 54 cu yds

Pavement Removal:

3" thick, (10.5' × 81')/9 = 95 sq yds

Concrete Removal:

(6"/12 × 10.5' × 81')/27 = 16 cu yds

Time Estimate

Concrete:

Roof — 29 cu yds × $\frac{1 \text{ day}}{10.7 \text{ cu yds}}$ = ⟶ 3 days

Mat — 37 cu yds × $\frac{1 \text{ day}}{20.3 \text{ cu yds}}$ = ⟶ 1 1/3 days

Walls — 46 cu yds ⎫
Vent — 28 cu yds ⎬ 74 cu yds × $\frac{1 \text{ day}}{11.2 \text{ cu yds}}$ = 7 days

Total = 11 1/3 days

Sheeting:

3080 sq ft. × $\frac{1 \text{ day}}{295 \text{ sq ft.}}$ = 10 1/2 days

---

Excavation:

609 cu yds × $\frac{1 \text{ day}}{480 \text{ cu yds}}$ = 1 1/3 days USE 2 days

---

Backfilling:

264 cu yds × $\frac{1 \text{ day}}{235 \text{ cu yds}}$ = 1 1/8 days USE 2 days

---

Stone Ballast:

26 cu yds × $\frac{1 \text{ day}}{160 \text{ cu yds}}$ = USE 1 day

Pavement Removal:

$$95 \text{ sq yds} \times 1.3 \times \frac{1 \text{ day}}{690 \text{ sq yds}} = \text{USE } \tfrac{1}{2} \text{ day}$$

Concrete Removal:

$$16 \text{ cu yds} \times \frac{1 \text{ day}}{45 \text{ cu yds}} = \text{USE } \tfrac{1}{2} \text{ day}$$

Roadway Decking:

12"×12" Timbers — $9818 \text{ B.F.} \times \dfrac{1 \text{ day}}{800 \text{ B.F.}} = 12\tfrac{1}{4} \text{ days}$

Steel —

$$2.125 \text{ tons} \times \frac{1 \text{ day}}{7.5 \text{ tons}} = \tfrac{1}{4} \text{ day}$$

Removal of Concrete Bulkhead:

$$28 \text{ cu yds} \times \frac{1 \text{ day}}{34 \text{ cu yds}} = \text{USE } 1 \text{ day}$$

Removal of Stone Wall:

$$300 \text{ cu ft.} \times \frac{1 \text{ day}}{900 \text{ cu ft.}} = \text{use } 1 \text{ day}$$

Tunnel Flooring:

4"×8" — $760 \text{ L.F.} \times \dfrac{1 \text{ day}}{160 \text{ L.F.}} = 4\tfrac{3}{4} \text{ days}$

plywood — $1{,}200 \text{ sq ft.} \times \dfrac{1 \text{ day}}{1{,}350 \text{ sq ft.}} = \text{use } 1 \text{ day}$

Excavate Existing Tunnel Fill:

$$195 \text{ cu yds} \times \frac{1 \text{ day}}{6 \text{ cu yds}} = 32\frac{1}{2} \text{ days}$$

Electrical & Plumbing Work:

$$.10 \times 81 \text{ days} = 8 \text{ days say } 10 \text{ days}$$

$$\text{Total} = 91 \text{ working days}$$

$$\frac{91 \text{ days}}{20 \text{ days/month}} = 4.55 \text{ months} \times 1.2 = 5.5 \text{ mos}$$
↑
contingency

Say 6 months

× 2 = 1 YEAR

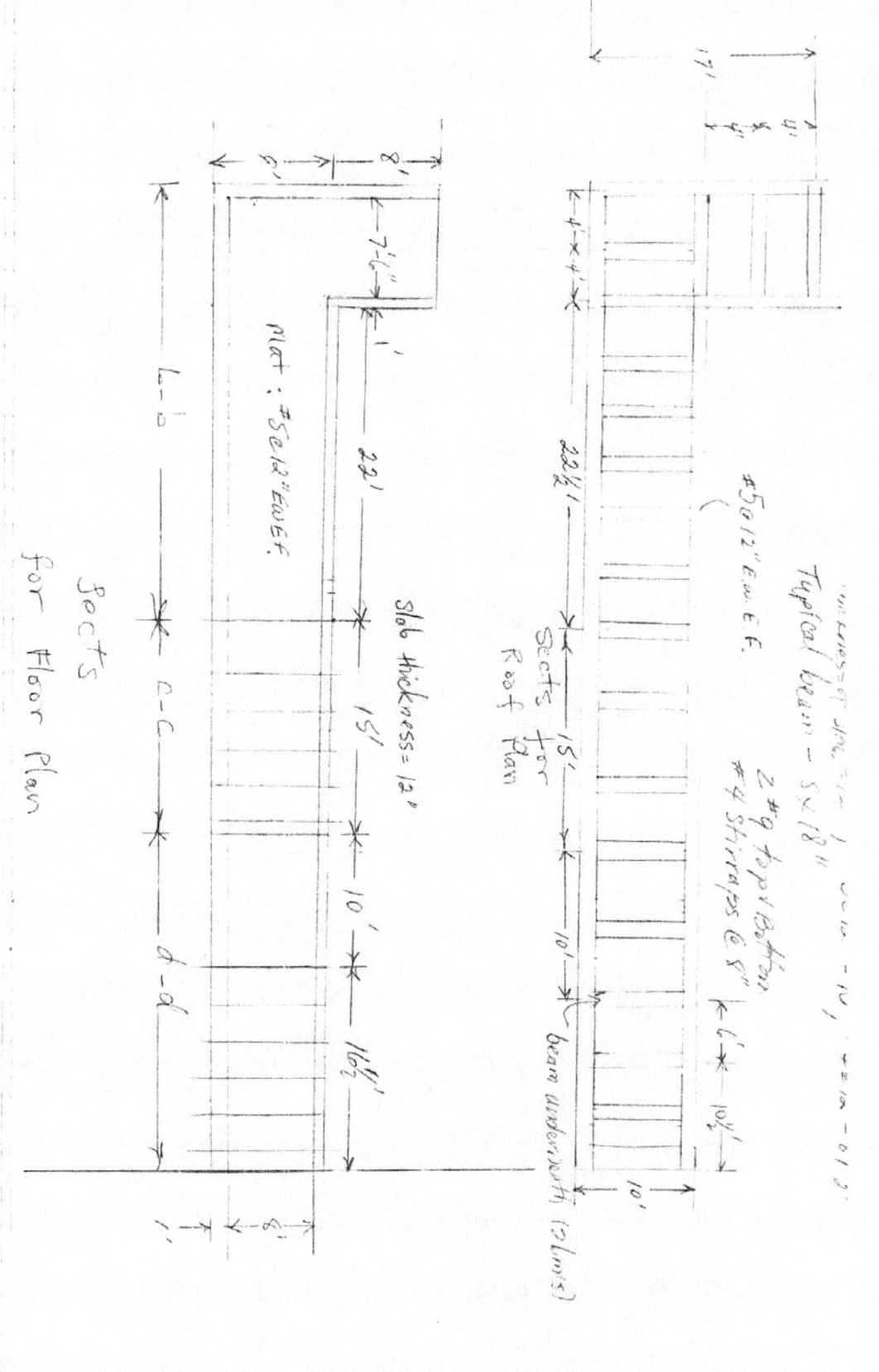

Sects for Floor Plan

1. $w = \frac{1}{2}$ k/ft @ 0' depth
2. $w = 1$ k/ft @ 10' depth
3. $w = 2$ k/ft @ 16' depth

$f'_c = 4000 \text{ psi}$

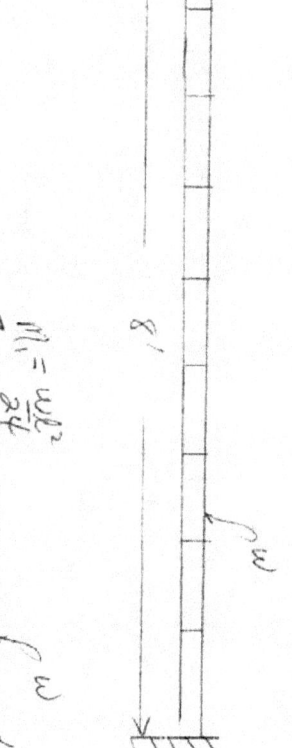

1. $w_t = \frac{1}{2}(1000/ft) \times 8' = 4000 \text{ lbs}$
2. $w_t = 1000 \text{ lbs/ft} \times 10' = 10,000 \text{ lbs}$
3. $w_t = 2000 \text{ lbs/ft} \times 16' = 32,000 \text{ lbs}$

1. $M_1 = .5 \text{k/ft} \times 8^2/24 = 1.33 \text{ k-ft}$, $M_2 = .5 \text{k/ft} \times 8^2/8 = 4 \text{ k-ft}$
2. $M_1 = 1 \text{k/ft} \times 8^2/24 = 2.67 \text{ k-ft}$, $M_2 = 1 \text{k/ft} \times 8^2/8 = 8 \text{ k-ft}$
3. $M_1 = 2 \text{k/ft} \times 8^2/24 = 5.33 \text{ k-ft}$, $M_2 = 2 \text{k/ft} \times 8^2/8 = 16 \text{ k-ft}$

$V_1 = 2\sqrt{f'_c} \cdot b \cdot d = 2\sqrt{4000} \times 8/12 \times 1' = 84 \text{ psi}$; $V_2 = 84 \text{ psi}$; $V_3 = 84 \text{ psi}$

Originally Handwritten by Stephen L. Carroll P.E.

Item	Description	Unit	Quantity	Price	Amount
1	Obligation	L.S.			$15,000
2	Sheeting	sq.ft.	7,260	28.51	$206,983
3	Decking	B.M.	22,303	1.59	$35,462
4	Earth Excavation	cu.yd.	1,476	7.55	$11,144
5	Excavation of Existing Tunnel Fill	cu.yd.	195	49	$9,575
6	Stone Ballast	cu.yd.	56	27.80	$1,577
7	Backfill	cu.yd.	650	6.38	$4,147
8	Concrete for Entrances	cu.yd.	255	381	$97,155
9	Concrete for V	cu.yd.	28	276	$7,728
10	Steel Reinforcing	lbs	28,454	0.65	Included
11	Sleeves in Concrete	Each	100	20	$2000
12	Tunnel Flooring	sq.ft.	10,000	2	$20,000
13	Hand Rail	L.F.	1000	16.21	$16,210
14	Railroad Track	L.F.	1,000	80	$80,000
15	Saving Excavation Tunnel Fill	Stone	Unknown	5	Unknown
16	Recondition Stone Block Walls	sq.ft.	1,268	270	$3,434
17	Entrance Sculpture	L.S.	2	5,000	$10,000
18	Restoration of Sidewalk	sq.ft.	840	3.49	$2,932
19	Restoration of Pavement	sq.yd.	134	725	$972
20	Concrete Excavation	cu.yd.	36	183	$6,588
21	Pavement Excavation	sq.yd.	215	4.90	$1,054
22	Remove Stone Block Wall	cu.ft	600	3.07	$1,842
23	Remove Concrete Bulkhead	cu.ft	1,512	3.02	$4,566
24	Entrance Handrails	L.F.	160	34.99	$5,598

Originally Handwritten by Stephen L. Carroll P.E.

Item	Description	Unit	Quantity	Price	Amount
25	Non-Slip Stair Treds	L.F.	608	12.97	$7,886
26	Entrance Hatch Doors	Each	2	7,000	$14,000
27	Vent Stacks @ Entrance	L.F.	56	89.30	$5,000
28	Steel Ladders @ Vents	lbs	6000	1	$6,000
29	3-ply Waterproofing	sq.ft.	1,692	1.69	$2,860
30	4-ply Waterproofing	sq.ft.	1,692	2.25	$3,807
31	Fire Stand Pipes	L.F.	1,663	34	$56,542
32	Exhaust Fans	Each	3	4,083	$12,249
33	Plumbing	L.S.			$28,000
34	Electrical	L.S.			$48,000
35	Replacement of Mains	L.S.			$118,000
36	Maintenance and Utilities	L.S.			$10,000
37	Public Address	L.S.			$3,000
38	Telephone	L.S.			$1,000
39	Crane to Lower Trolley	L.S.			$1,000
40	Refreshments, Souvenirs, & Exhibits	L.S.			$7,000
41	Borings and Test Pits	L.S.			$3,446
42	Bond @1%	L.S.			$8,804
43	Contingency (20%)	L.S.			$176,078
44	Administrative and Engineering Fees	L.S.			$125,000
45	Exterminator				
Total					$1,190,272
Say					$1,200,000

Updated Cost Estimates

1985 Cost Estimates for **One** Tunnel Entrance: $1,200,000

1985 Construction Cost Index (1): 5388.08

July 2015 Construction Cost Index (2): 10037.40

Inflation Factor from 1985 to 2015

10037.40/5388.08 = 1.86

Updated 2015 Cost:

1,200,000 x 1.86 = $2,235,412

Cost Per Each Entrance: **$2,235,412**

Updated Cost Estimate Using Consumer Price Index (CPI) Cumulative Rate of Inflation: 121.8% (3)

1985 Cost: $1,200,000

2015 Cost: **$2,661,568.77**

Note: Use the higher figure

(1) From Engineering News Record

(2) From Engineering News Record

(3) http://www.usinflationcalculator.com/

Inflating Construction Costs to 2015 Dollars

City Cost Index - New York - As of November 2011

The building and construction cost indexes for ENR's individual cities use the same components and weighting as those for the 20-city national indexes. The city indexes use local prices for portland cement and 2 X 4 lumber and the national average price for structural steel. The city's **BCI** uses local union wages, plus fringes, for carpenters, bricklayers and iron workers. The city's **CCI** uses the same union wages for laborers.

ENR COST INDEXES IN NEW YORK (1978-2011)

YEAR	MONTH	BCI	%CHG	CCI	%CHG
2011	Nov	7242.23	3.4	13807.28	3.0
2011	Oct	7237.23	3.5	13802.28	3.0
2011	Sep	7230.73	3.6	13795.78	3.1
2011	Aug	7229.48	3.2	13794.53	2.9
2011	Jul	7216.73	3.0	13781.78	2.8
2011	Jun	7205.23	2.9	13770.28	2.7
2011	May	7039.93	0.8	13441.53	0.4
2011	Apr	7038.93	3.2	13440.53	3.8
2011	Mar	7029.43	3	13431.03	3.7
2011	Feb	7026.43	3.1	13428.03	3.7
2011	Jan	6994.68	2.7	13396.28	3.5
2010	Dec	6998.68	2.7	13400.28	3.5
2010	Nov	7001.93	2.8	13403.53	3.6
2010	Oct	6993.43	2.7	13395.03	3.5
2010	Sep	6982.93	2.4	13384.53	3.4
2010	Aug	7008.52	2.8	13410.12	3.6
2010	Jul	7008.02	2.8	13409.62	3.6
2010	Jun	7001.52	2.5	13403.12	3.4
2010	May	6984.77	2.1	13386.37	3.2
2010	Apr	6824.18	-0.4	12951.62	-0.2
2010	Mar	6822.18	-0.4	12949.62	-0.2
2010	Feb	6816.18	-0.5	12943.62	-0.3
2010	Jan	6814.43	-0.7	12941.87	-0.4
2009	Dec	6816.18	-0.9	12943.62	-0.5
2009	Nov	6813.43	-1.7	12940.87	-0.9
2009	Oct	6813.18	-1.9	12940.62	-1
2009	Sep	6817.93	-1.9	12945.37	-1
2009	Aug	6818.68	-1.2	12946.12	-0.6
2009	Jul	6816.93	-1.1	12944.37	-0.6
2009	Jun	6829.68	2.6	12957.12	3.5
2009	May	6842.43	3.5	12969.87	3.9

Year	Month				
2009	Apr	6849.93	3.9	12977.37	4.1
2009	Mar	6851.68	4	12979.12	4.2
2009	Feb	6849.18	4.2	12976.62	4.3
2009	Jan	6863.43	4.4	12990.87	4.4
2008	Dec	6875.68	4.6	13003.12	4.5
2008	Nov	6928.68	5.5	13056.12	5
2008	Oct	6947.43	5.6	13074.87	5.1
2008	Sep	6952.68	5.7	13080.12	5.1
2008	Aug	6900.93	4.9	13028.37	4.7
2008	Jul	6894.18	4.9	13021.62	4.7
2008	Jun	6855.53	1.3	12523.87	0.7
2008	May	6613.53	0.7	12481.87	0.4
2008	Apr	6594.28	0.6	12462.62	0.3
2008	Mar	6589.28	1.1	12457.62	0.6
2008	Feb	6573.28	0.9	12441.62	0.5
2008	Jan	6573.28	0.9	12441.62	0.5
2007	Dec	6571.53	0.8	12439.87	0.4
2007	Nov	6569.03	0.5	12437.37	0.3
2007	Oct	6578.03	1	12446.37	0.6
2007	Sep	6577.03	3.6	12445.37	3.3
2007	Aug	6576.78	3.7	12445.12	3.4
2007	Jul	6574.78	3.7	12443.12	3.4
2007	Jun	6568.28	3.3	12436.62	3.1
2007	May	6567.28	4.3	12435.62	3.7
2007	Apr	6553.28	4.1	12421.62	3.6
2007	Mar	6515.56	3.6	12383.9	3.3
2007	Feb	6513.06	3.4	12381.4	3.2
2007	Jan	6513.56	3.2	12381.9	3.1
2006	Dec	6520.06	3.4	12388.4	3.2
2006	Nov	6535.31	4	12403.65	3.6
2006	Oct	6510.56	4.4	12378.9	3.7
2006	Sep	6349.46	2.5	12045.4	1.3
2006	Aug	6347.21	2.6	12043.15	1.3
2006	Jul	6340.21	2.8	12036.15	3.2
2006	Jun	6361.79	2	12057.73	2.6
2006	May	6298.56	1.1	11994.49	2.3
2006	Apr	6293.54	1.9	11989.47	2.2
2006	Mar	6290.54	2.2	11986.47	2.4
2006	Feb	6296.54	1.9	11992.47	2.2
2006	Jan	6310.51	1.9	12006.44	2.2
2005	Dec	6304.51	3.2	12000.44	2.9
2005	Nov	6281.61	2.8	11977.54	2.7
2005	Oct	6239.32	2	11935.25	2.3
2005	Sep	6194.31	0.9	11890.24	1.7
2005	Aug	6188.06	5	11883.99	7
2005	Jul	6169.29	3.3	11667.99	4.3
2005	Jun	6235.19	4.7	11733.89	5
2005	May	6230.38	4.9	11729.08	5.1
2005	Apr	6179.09	5.3	11729.08	5.8
2005	Mar	6154.37	6.2	11704.36	6.3
2005	Feb	6179.22	7.1	11729.21	6.8
2005	Jan	6193.86	8.7	11743.85	7.2
2004	Dec	6112.26	9.5	11662.25	12.3
2003	Dec	5583.09	2.7	10386.73	3.8
2002	Dec	5438.2	2	10009.06	-0.9

Year	Month					
2001	Dec	5330.03	6.2	10101.24	7.7
2000	Dec	5018.67	-2.5	9379.14	0.3
1999	Dec	5147.21	5.3	9355.77	5.1
1998	Dec	4890.13	0.2	8899.59	1.8
1997	Dec	4880.61	2.2	8742.88	2.2
1996	Dec	4774.23	4.8	8554.47	2.1
1995	Dec	4557.44	2.2	8378.68	3.2
1994	Dec	4458.36	2.5	8117.64	4.9
1993	Dec	4349.2	4.8	7737.11	5
1992	Dec	4151.28	3.8	7367.49	3.6
1991	Dec	3997.91	3.9	7110.37	3.9
1990	Dec	3847.21	3.6	6846.49	6.1
1989	Dec	3712.2	5.4	6453.56	3.6
1988	Dec	3522.07	4.5	6231.12	4.5
1987	Dec	3369.28	4.7	5961.27	6.1
1986	Dec	3217.83	4.5	5621.15	4.3
1985	Dec	3076.19	3.1	5388.08	4.4
1984	Dec	2983.27	6.8	5160.95	5.6
1983	Dec	2792.67	7.3	4887.55	7.3
1982	Dec	2603.28	6.9	4553.93	10.4
1981	Dec	2434.62	11.3	4125.68	9.3
1980	Dec	2188.06	4.6	3774.64	5.4
1979	Dec	2091.82	11.5	3580.5	7.7
1978	Dec	1875.62	2	3325.43	7.4

Subscribe to ENR | Back Issues | Manage your subscription | Get Top List Plaques

Reader Comments:

Add a comment

[Submit]

POWERED BY Pluck

View This Week's Magazine | Go to Magazine Archives | Subscribe to ENR | Order Back Issues

Marketplace Links

Basics of a Mixed Domain Oscilloscope
Discover the benefits of seeing both the time & frequency domains in one glance

i This site uses cookies. By continuing to browse the site you are agreeing to our use of cookies. Review our **Privacy and Cookie Notice** for more details.

subscribe | contact us | advertise | industry jobs | events | FAQ | ENR Subscriber Login »

SUBSCRIBE TODAY
Web access will be provided
as part of your subscription.

Search our site: Enter your search...

| INFRASTRUCTURE | BLDGS | BIZ MGMT | POLICY | EQUIPMENT | PEOPLE | MULTIMEDIA | OPINION | TECH | EDUCATION | **ECONOMICS** | TOP LISTS | REGIONS |

CURRENT COSTS | MATERIAL TRENDS | HISTORICAL INDICES | QUARTERLY COST REPORTS | FAQ

Construction Economics

ENR publishes both a Construction Cost Index and Building Cost index that are widely used in the construction industry. This website contains an explanation of the indexes methodology and a complete history of the 20-city national average for the CCI and BCI. Both indexes have a materials and labor component. In the second issue of each month ENR publishes the CCI, BCI, materials index, skilled labor index and common labor index for 20 cities and the national average. The first issue also contains an index review of all five national indexes for the latest 14 month period.

Click here for more on using ENR Indexes

Current Cost Indices

| Construction Costs | Building Costs | Materials Cost |

+2.1% CONSTRUCTION COST
Jul 2015

1913 = 100	INDEX VALUE	MONTH	YEAR
CONSTRUCTION	10037.4	0.0%	+2.1%
COMMON LABOR	21463.03	0.0%	+2.2%
WAGE $/HR.	40.78	0.0%	+2.9%

Gain Quick Access to ENR's Historical Cost Index Tables

Find ENR's Award-Winning Building Material Price Data (published later than March 1, 2005). Monthly tables on 75 different building materials! Click here.

Dodge Lead Center

Search for local construction projects OR CALL **877-234-4246** and get a FREE Lead Now!

Search by Project Type & State

All Project Types

Select a State (Required)

SEARCH

Materials Trends

View all **Materials Trends »**

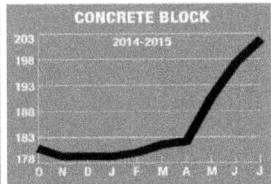

ENR's asphalt, cement and concrete prices for July 2015.

Portland cement prices drop
Portland cement prices dropped 0.1% this month, per ENR's 20-city average, after seeing a 0.3% increase in June. According to the Bureau of Labor Statistics' producer price index, the price for cement did not change from April to May after increasing 2.8% the previous month but saw a boost of 8.4% since the same time last year. Ready-mix concrete prices are up 3.3% over July 2014, declining slightly from an annual escalation rate of 3.5% in June. The price of concrete block held steady since last month but is up 2.2% annually, dropping from June's 3.7% yearly increase rate.

Click for full report [July 06, 2015]

Historical Indices

View all **Historical Indices »**

▶ **Construction Cost Index History**
200 hours of common labor at the 20-city average of common labor rates, plus 25 cwt of standard structural steel shapes at the mill price prior to 1996 and the fabricated 20-city price from 1996, plus 1.128 tons of portland cement at the 20-city price, plus 1,088 board-ft of 2 x 4 lumber at the 20-city price.

▶ **Building Cost Index History**
68.38 hours of skilled labor at the 20-city average of bricklayers, carpenters and structural ironworkers rates, plus 25 cwt of standard structural steel shapes at the mill price prior to 1996 and the fabricated 20-city price from 1996, plus 1.128 tons of portland cement at the 20-city price, plus 1,088 board ft of 2 x 4 lumber at the 20-city price.

Cost Index History Tables

The building and construction cost indexes for ENR's individual cities use the same components and weighting as those for the 20-city national indexes. The city indexes use local prices for portland cement and 2 X 4 lumber and the national average price for structural steel. The city's BCI uses local union wages, plus fringes, for carpenters, bricklayers and iron workers. The city's CCI uses the same union wages for laborers.

ENR Cost Indexes in 20 Cities 1978-2012

▶ Atlanta, GA	▶ Baltimore, MD	▶ Birmingham, AL	▶ Boston, MA
▶ Chicago, IL	▶ Cincinnati, OH	▶ Cleveland, OH	▶ Dallas, TX
▶ Denver, CO	▶ Detroit, MI	▶ Kansas City, MO	▶ Los Angeles, CA
▶ Minneapolis, MN	▶ New Orleans, LA	▶ New York, NY	▶ Philadelphia, PA
▶ Pittsburgh, PA	▶ San Francisco, CA	▶ Seattle, WA	▶ St. Louis, MO

ENR Cost Indexes in Canadian Cities 1978-2012

| ▶ Montreal | ▶ Toronto |

The US Inflation Calculator measures the buying power of the dollar over time. Just enter any two dates between 1913 and 2015, an amount, and click 'Calculate'.

Inflation Calculator

If in [1985] (enter year)

I purchased an item for $ [1,200,000.00]

then in [2015] (enter year*)

that same **item would cost**: $2,661,568.77

Cumulative rate of inflation: 121.8%

CALCULATE   PRINT

*Learn how this calculator works. This US Inflation Calculator uses the latest US government CPI data published on August 19, 2015 to adjust for inflation and calculate the cumulative inflation rate through July 2015. The Consumer Price Index (CPI) and inflation for August 2015 is scheduled for release by the United States government on September 16, 2015. (See a chart of recent inflation rates.)

INFLATION

US Inflation in July Edges Up 0.1%; Annual Inflation Rate at 0.2%

AUGUST 19, 2015 | LEAVE A COMMENT

US inflation rose for a sixth straight month in July but at a markedly slower pace than in June and May, according to government data released on Wednesday, August 19.

American consumers spent less on vehicles, transportation and some energy items like fuel oil and electricity but they continued to pay more for food, gasoline and housing. In addition, clothing and healthcare costs went up.

Engineer's Report on Buried Steam Locomotive

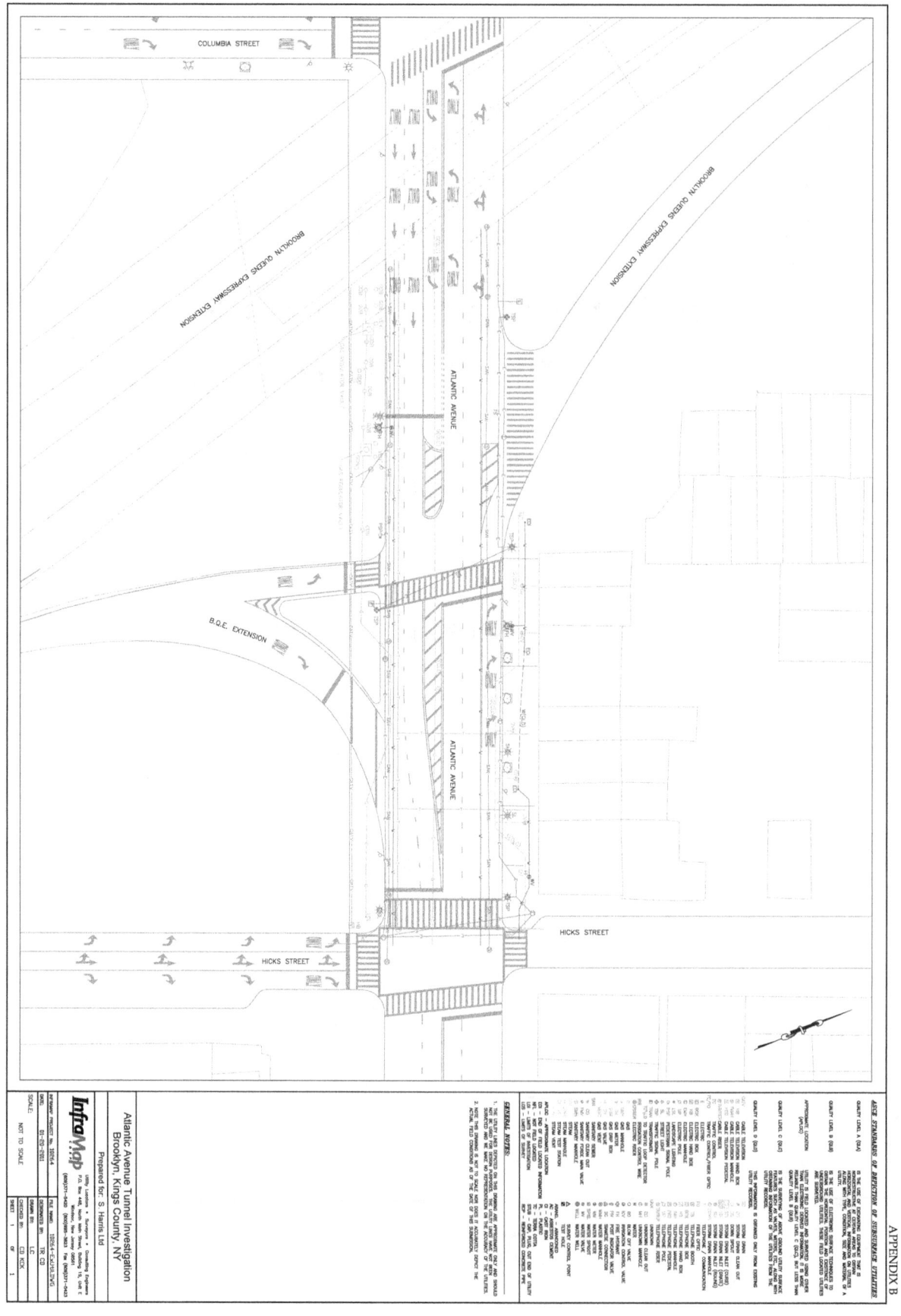

APPENDIX B

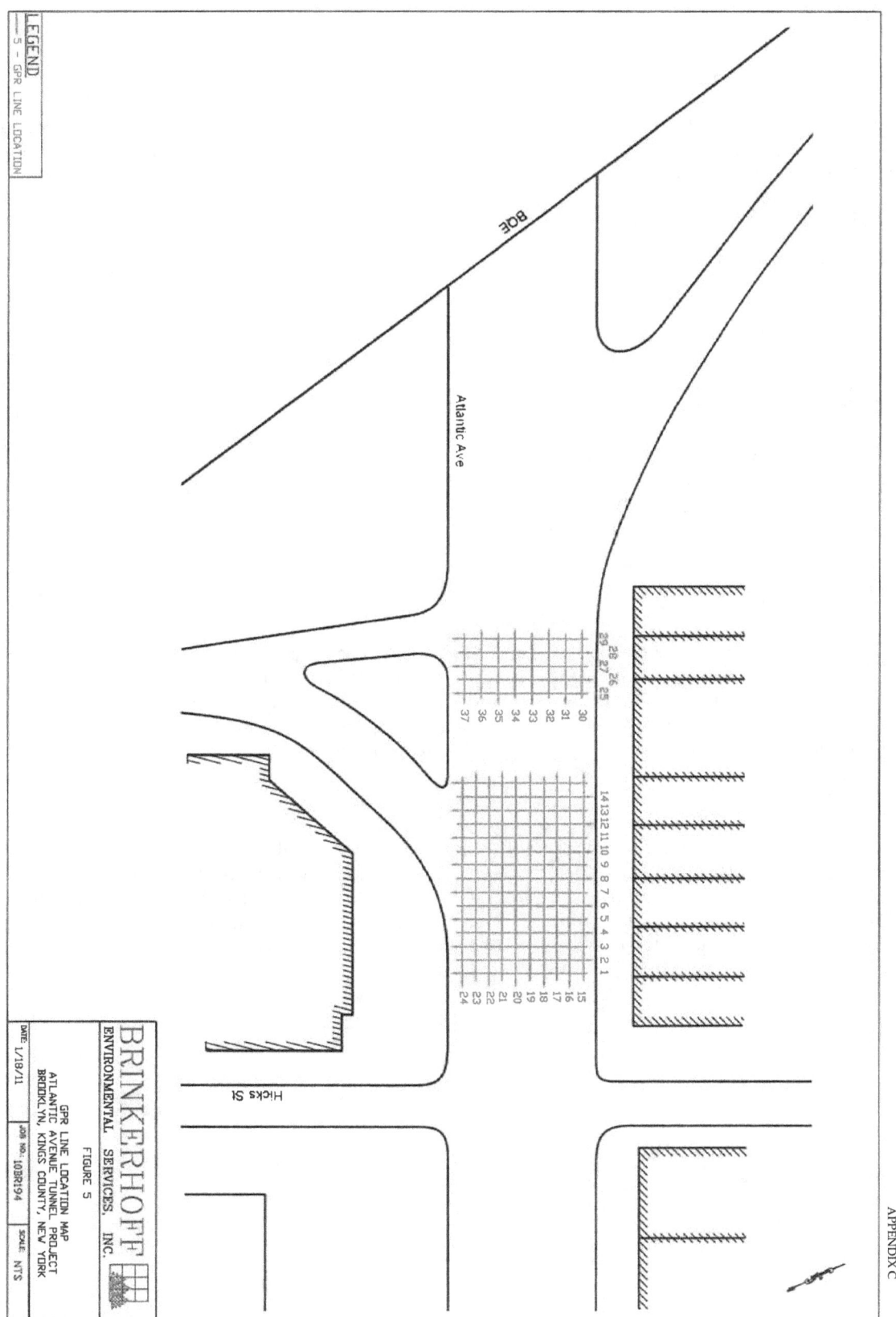

APPENDIX C

BRINKERHOFF
ENVIRONMENTAL SERVICES, INC.

1913 Atlantic Avenue, Suite R5
Manasquan, New Jersey 08736
Tel: (732) 223-2225
Fax: (732) 223-3666

January 18, 2011

Janine Hildebrand, EIT
S. Harris, Ltd.
2601 Pennsylvania Avenue, Suite Eight
Philadelphia, PA 19130

Re: Geophysical Investigation Report
     Atlantic Avenue
     Brooklyn, New York
     Brinkerhoff Project No. 10BR194

Dear Ms. Hildebrand:

Brinkerhoff Environmental Services, Inc. (Brinkerhoff) is pleased to present the following summary report of the Geophysical Investigation conducted on January 11 and 14, 2011 at the above-referenced property (herein referred to as the subject property). Refer to Figure 1 – Site Location Map. Electromagnetic induction, electromagnetic soil conductivity, total field magnetics and ground penetrating radar (GPR) were employed for the investigation.

Introduction

On January 11 and 14, 2011, Brinkerhoff conducted a geophysical investigation on the subject property. The purpose of the geophysical investigation was to evaluate the potential presence of subsurface anomalies indicative of a buried 19th century locomotive and associated artifacts. The subject property is currently an active urban roadway surrounded by buildings to the north, south and east, while a large steel overpass (I-278) borders the subject property to the west.

APPENDIX C

Janine Hildebrand, EIT
Re: Geophysical Investigation Report
    Atlantic Avenue
    Brooklyn, New York
    Brinkerhoff Project No. 10BR194
January 18, 2011

Methodology and Limitations

A. ***Electromagnetic Induction*** – Electromagnetic data were collected using a Geonics EM61-MK2A High Sensitivity Metal detector (EM-61). The EM-61 was operated in the differential mode while recording magnetic metallic response measurements. The effective depth of data collection was approximately 10 feet. The field procedure involved device calibration, data collection and recording, and data storage for analysis in the office. Data were recorded on the Allegro Field Computer. Magnetic data and Differential Global Positioning System (DGPS) data, were simultaneously recorded at zero-point-two (0.2) second intervals (approximately every zero point five [0.5] feet) along survey lines at approximately two-point-five (2.5)-foot intervals. The data were downloaded to a personal computer for processing and the creation of an EM61 response contour map. Refer to Figure 2 – EM-61 Data Map

B. ***Electromagnetic Soil Conductivity Survey*** – Electromagnetic data were collected using a Geonics EM-31 Terrain Conductivity Meter. The EM-31 was operated in the vertical dipole mode while recording ground conductivity measurements. The effective depth of data collection was approximately 15 feet. The field procedures involved device calibration, data collection and recording, and data storage for analysis in the office. Data were recorded on the Allegro Filed Computer. Conductivity data and DGPS data, were recorded at zero-point-two (0.2) second intervals (approximately every zero point five [0.5] feet) along survey lines at approximately five (5)-foot intervals. The data were downloaded to a personal computer for processing and the creation of a conductivity contour map. Refer to Figure 3 – EM-31 Data Map.

C. ***Total Field Magnetics*** - The G-859 Self-oscillating split-beam Cesium Vapor Magnetometer (G-859) was operated in the simple survey mode while recording magnetic metallic response measurements. The effective depth of data collection was approximately 20 feet. The field procedure involved device calibration, data collection and recording, and data storage for analysis in the office. Data were collected in two (2) separate survey files, recorded on the G-859 console data logger and transferred via high speed USB cable to a portable computer for further analysis and map generation within the field. Magnetic data and integrated Wide Area Augmentation System (WAAS) GPS data were simultaneously collected throughout the total field magnetic survey with survey lines spaced approximately 10 feet apart. The data were downloaded to a personal computer for processing and the creation of a G-859 response contour map. Refer to Figure 4 – G-859 Data Map

Janine Hildebrand, EIT
Re: Geophysical Investigation Report
     Atlantic Avenue
     Brooklyn, New York
     Brinkerhoff Project No. 10BR194
January 18, 2011

D. Ground-Penetrating Radar (GPR) Survey - GPR data were collected with a Sensors and Software Inc. Noggin^{plus} SmartCart GPR System (SmartCart) utilizing a 250 MHz antenna. Data were collected continuously on 38 survey lines across selected open areas of the subject property. The survey lines were spaced approximately two (2) feet apart and oriented perpendicular to each other. The depth of investigation was from zero (0) to approximately three-point-five (3.5) feet with this antenna. The data were processed using Ekko View Deluxe software. Refer to Figure 5 – GPR Line Map.

E. Limitations - Limitations encountered during the investigation included the presence of possible subsurface utilities, metallic light poles, suspect concrete road bedding, vehicles, vegetation, snow, refuse piles, adjacent structures and the I-278 overpass. Please note that Electromagnetic Induction, Terrain Conductivity, Total Field Magnetics and GPR measurement are remote sensing methods and in some instances, due to interference or other geophysical limitations, do not reveal data which may be indicative of subsurface anomalies. The findings of this investigation should only be used as a tool in evaluating the possibility that a locomotive is present on the property and should not be considered a guarantee regarding the presence or absence of a locomotive.

Geophysical Results

EM-61 Results: The EM-61 survey was limited to all outside accessible areas of the subject property. Several areas of anomalous change in magnetic susceptibility gradient were seen in the EM-61 data. Analysis of the EM-61 data showed that these anomalies coincided with observable surface features and/or the location of possible building materials.

EM-31 Results: The EM-31 survey was limited to all outside accessible areas of the subject property. Several areas of anomalous change in magnetic susceptibility gradient were seen in the EM-31 data. Analysis of the EM-31 data showed that these anomalies coincided with observable surface features and/or the location of possible building materials. One (1) large anomaly was identified within the EM-31 data and the location of the anomaly is shown on Figure 3. Brinkerhoff then further investigated anomaly A-1 with GPR.

G-859 Results: G-859 survey was limited to all outside accessible areas of the subject property. One (1) area of anomalous change in magnetic susceptibility gradient was seen in the G-859 data. Analysis of the G-859 data revealed a large metallic anomaly measuring approximately 20 feet in length. The location of the anomaly is shown on Figure 4. Brinkerhoff then further investigated anomaly A-1 with GPR.

APPENDIX C

Janine Hildebrand, EIT
Re: Geophysical Investigation Report
Atlantic Avenue
Brooklyn, New York
Brinkerhoff Project No. 10BR194
January 18, 2011
Page 4 of 5

GPR Results: GPR data was collected from the areas of anomaly A-1. Due to the assumed unconsolidated geology, brick and assumed concrete below the surface of the roadway, GPR was unable to penetrate further then three-point-five (3.5) feet below grade. Brinkerhoff was unable to verify the presence of the large magnetic anomaly which was detected in both the EM-31 and G-859 surveys. Representative GPR profiles are presented below.

Anomaly A-1 – GPR data was collected from the area of Anomaly A-1, as noted in the EM-31 and G-859 data. Based upon the EM-31 and G-859 data images, the anomaly is located largely on the eastern side of Atlantic Avenue; however; the anomaly's large response extends across Atlantic Avenue encompassing the western lanes as well. GPR data collected in the area of A-1 is inconclusive due to restricted GPR signal penetration within the subsurface geology. A representative GPR profile collected from this area showing A-1 and the GPR's restricted signal is shown below.

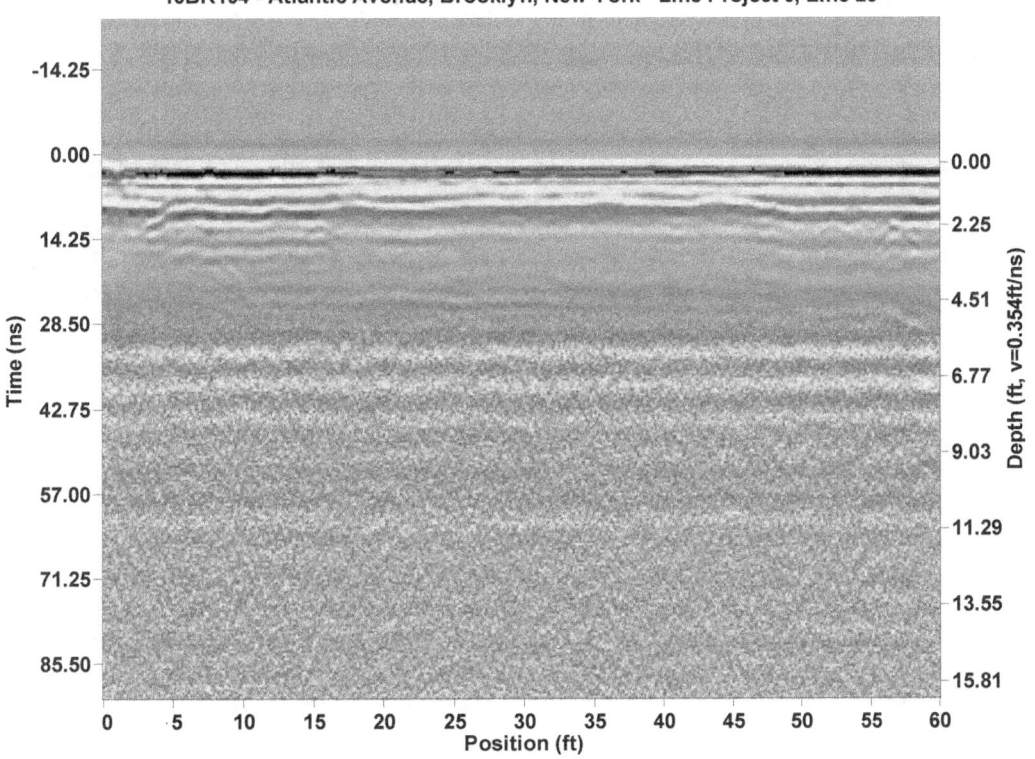

APPENDIX C

Janine Hildebrand, EIT
Re: Geophysical Investigation Report
      Atlantic Avenue
      Brooklyn, New York
      Brinkerhoff Project No. 10BR194
January 18, 2011

GEOPHYSICAL CONCLUSIONS

On January 11 and 14, 2011, Brinkerhoff performed a geophysical investigation in open and accessible areas of the subject property. Limitations encountered during the investigation included the presence of possible subsurface utilities, metallic light poles, suspect concrete road bedding, vehicles, vegetation, snow, refuse piles, adjacent structures and the I-278 overpass. Please note that Electromagnetic Induction, Terrain Conductivity, Total Field Magnetics and GPR measurement are remote sensing methods and in some instances, due to interference or other geophysical limitations, do not reveal data which may be indicative of subsurface anomalies. One (1) large subsurface metallic anomaly was identified extending across Atlantic Avenue and encompassing both the west bound and east bound roadway. Brinkerhoff was able to estimate the metallic anomaly's length at 20 feet based upon the response of the G-859 data. The anomaly was outlined in white spray paint in the field and is noted on the attached Figure 3 and Figure 4.

This report has been prepared and is respectfully submitted by

BRINKERHOFF ENVIRONMENTAL SERVICES, INC.

January 18, 2011
_____   _____
MATTHEW D. POWERS   Date
Director of Geophysical Services

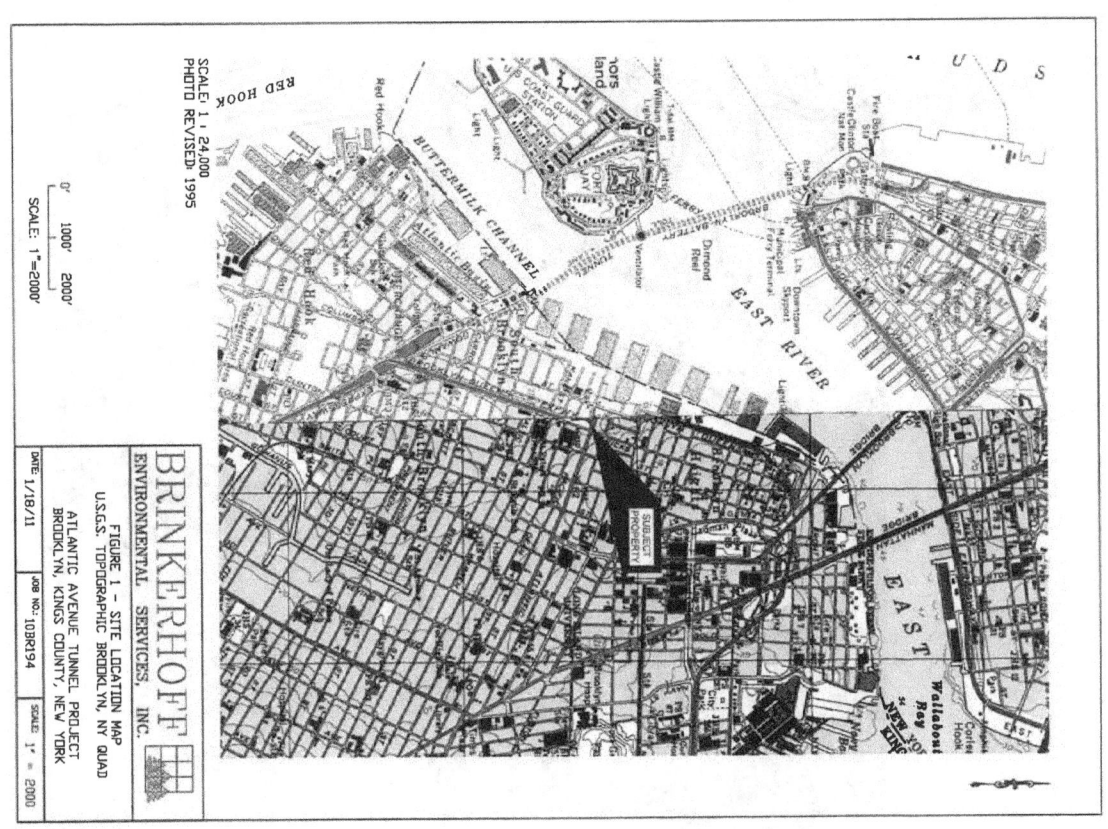

Figure 2
Aerial Photograph with EM-61 Data
Atlantic Avenue Tunnel Project
Brooklyn, Kings County, New York

Figure 3
Aerial Photograph with EM-31 Data
Atlantic Avenue Tunnel Project
Brooklyn, Kings County, New York

Figure 4
Aerial Photograph with G-859 Data
Atlantic Avenue Tunnel Project
Brooklyn, Kings County, New York

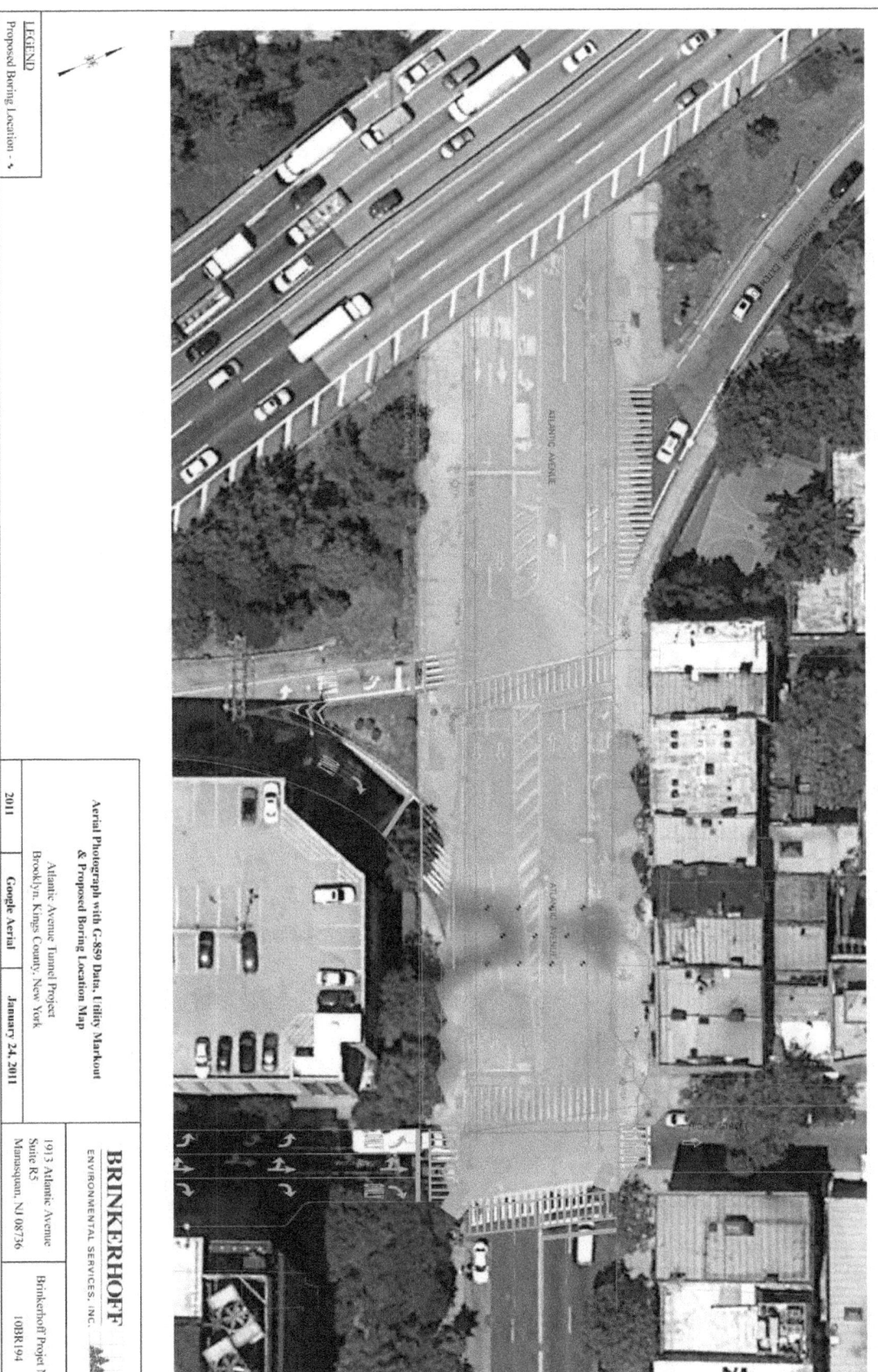

Defibrillator and Related Equipment

Home > Business AED Packages >

ZOLL AED Plus Business AED Package

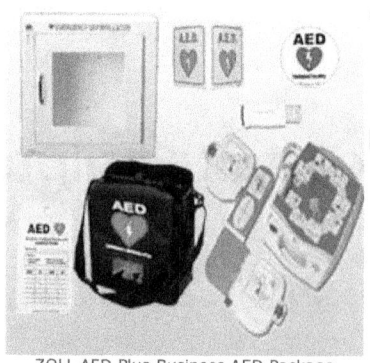

ZOLL AED Plus Business AED Package

Our Price: $1,745.00

Availability:: Usually Ships in 24 to 48 Hours
Product Code: ZOLLBUSPKG

Description

ZOLL AED Plus Business Package
Includes-
    ZOLL AED Plus
    Adult CPR D Pads (5 year life)
    ZOLL AED Plus Batteries (5 year life)
    Set-Up Video
    Operators Manual
    5 year warranty
    Softcase
    Infrared Device USB (for updating software)
    Alarmed AED Cabinet
    AED Inspection Tag
    "AED Equipped Facility" Sticker
    V-Shaped AED Sign
    Free Shipping within the US

ZOLL AED Plus
When a cardiac arrest occurs, the fact is that only half of the victims will need a shock. The other half requires CPR. You deserve an AED that helps you all the time. And, only one AED can actually see when you are doing CPR and help you do it well. You need more than just commands, without assistance. That's not smart, and it's certainly not help.

Real CPR Help- The ZOLL AED Plus features Real CPR Help™, a tool that is able to actually see what you are doing and provide feedback to help you do it well. Audio and visual prompts help you rescue with confidence and clarity unmatched by any other automated external defibrillator (AED).

- Not pushing hard enough? The ZOLL AED Plus will tell you when to push harder.
- Pushing hard enough? The ZOLL AED Plus will say, "Good compressions."
- Not pushing fast enough? A metronome will lead you to the right rate.
- It will even show you the depth of each compression. In real time.
- Not yet started? The ZOLL AED Plus will tell you again to get started.
- Compressions stopped? The ZOLL AED Plus will tell you to continue.

Lowest Total Cost of Ownership over 10 years- ZOLL AED Plus Defibrillators feature 5 year pads and 5 year batteries. Total cost for 10 years is $204 and you only replace the batteries and pads once (in 10 years).

Rugged- The ZOLL AED Plus performs even when wet. The ZOLL AED Plus features an IP Rating of IP55 one of the highest on the market today. Read about the ZOLL AED Plus and water ratings.

ZOLL believes an AED should not just deliver a shock. It should also help the rescuer provide high-quality cardiopulmonary resuscitation (CPR). That's why you need the ZOLL AED Plus with Real CPR Help.

Resources
ZOLL AED Plus Manual
ZOLL AED Plus Product Brochure
ZOLL AED Plus Technical Specifications
ZOLL AED Plus Training and Accessories Brochure
ZOLL Ingress Protection (IP) Study

Confidence that comes only from Real CPR Help.

AED Plus
Step-by-step support for the entire Chain of Survival

The ZOLL AED Plus gives even infrequent rescuers the confidence they need to help save lives. Simple audio prompts and illustrations reinforce every step in the resuscitation process.

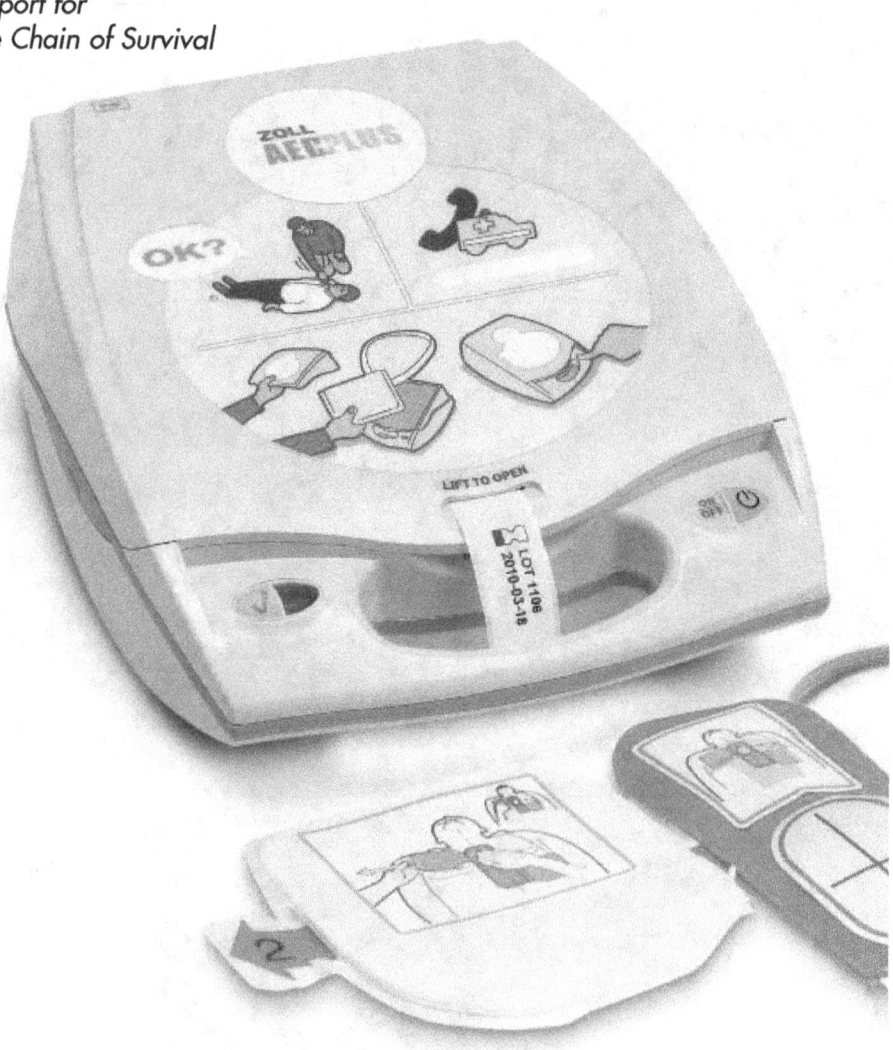

One-piece CPR-D-padz with Real CPR Help technology makes placement of electrodes faster and more accurate.

"Push Harder." "Good Compressions." "Press Shock Button."

Want to protect your employees and provide added peace of mind? Put ZOLL® automated external defibrillators (AEDs) in your workplace. As the brand hospitals choose most for professional defibrillators, we've earned the respect of users at every skill level, including infrequent rescuers. That's because our AEDs do more than provide the most advanced defibrillation technology. The AED Plus® and the AED Pro® are the only automated external defibrillators to provide Real CPR Help® with real-time feedback on the depth and rate of chest compressions during cardiopulmonary resuscitation (CPR). In addition to audio prompts, both the AED Plus and AED Pro feature on-screen instructions, and a bar gauge that shows how deep to push in order to optimize compression depth. This is an important advancement, given the American Heart Association Guidelines and their recommendation for more frequent, higher quality compressions during CPR.

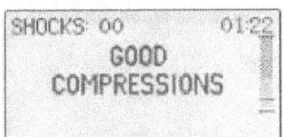

AED Pro
A rugged, advanced AED for professional users

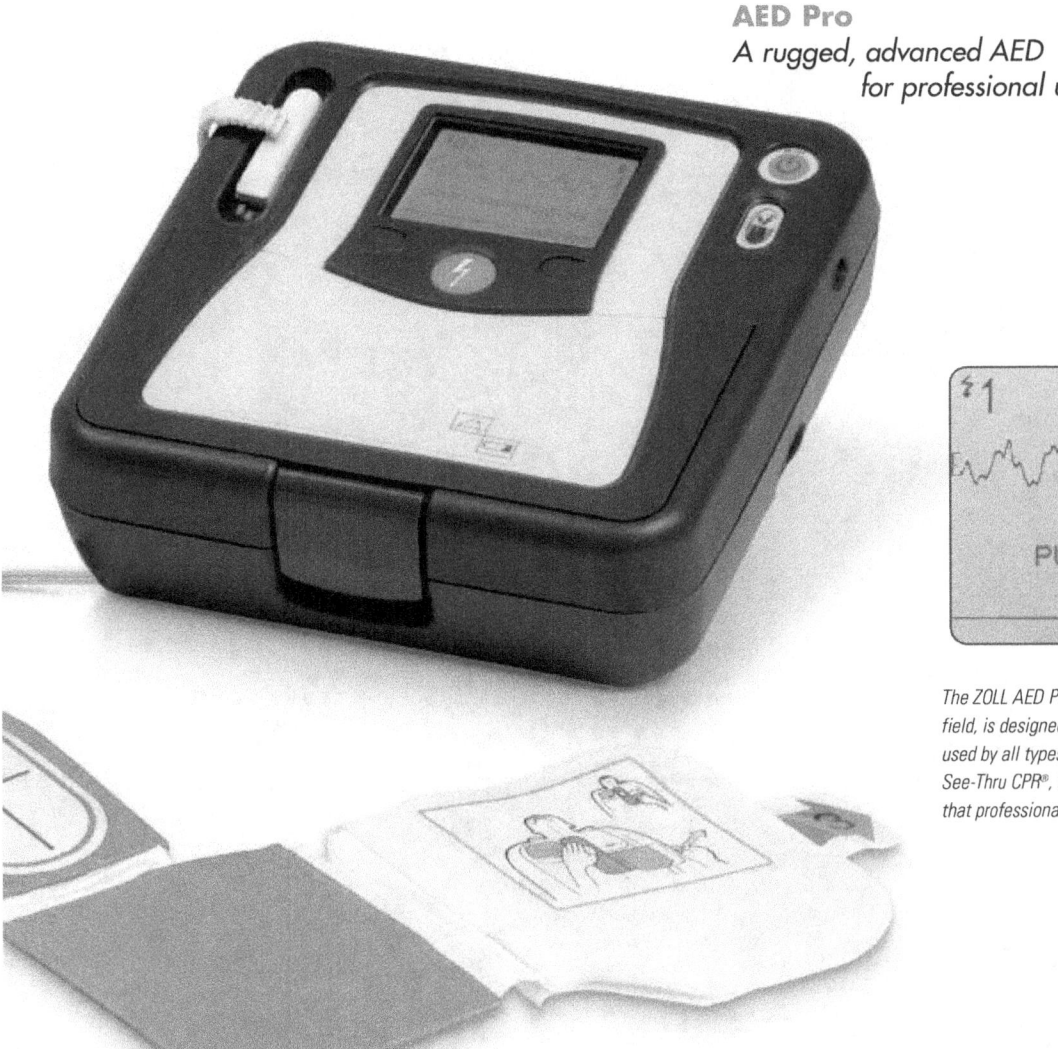

Chest compression depth gauge

The ZOLL AED Pro, whether used in a hospital or in the field, is designed to perform in any environment, and is used by all types of rescuers. With Real CPR Help and See-Thru CPR®, it provides the advanced functionality that professional rescuers and services require.

AED Plus

- Real CPR Help provides real-time feedback for depth and rate of CPR chest compressions
- Supports the entire Chain of Survival, with easy-to-understand audio prompts and illustrations
- Has lowest total cost of ownership, once installed, because CPR-D-padz and batteries last five years, if not used
- Rugged design that resists dust and direct water spray (IP55 rating)
- Powered by 10 lithium batteries available from retail stores everywhere

AED Pro

- Real CPR Help provides real-time feedback for depth and rate of CPR chest compressions
- See-Thru CPR filters CPR artifact to display patient's underlying ECG rhythm
- Advanced capabilities, including 3-lead patient monitoring and manual override
- The only AED that has an IP55 rating for dust and water resistance and passes the 1.5-meter drop test
- Battery and electrode compatibility with ZOLL's line of professional defibrillators

---

ZOLL Medical Corporation develops and markets medical devices and related software solutions that help advance emergency care and save lives, while increasing clinical and operational efficiencies. With products for defibrillation and monitoring, circulation and CPR feedback, data management, fluid resuscitation, and therapeutic temperature management, ZOLL provides a comprehensive set of technologies which help clinicians, EMS and fire professionals, and lay rescuers treat victims needing resuscitation and critical care.

A NASDAQ Global Select company and a Forbes 100 Most Trustworthy Company in 2007, 2008 and 2009, and a Forbes Top 100 Small Business Company in 2011, ZOLL develops and manufactures its products in the United States, in California, Colorado, Illinois, Massachusetts, Pennsylvania, and Rhode Island. More than 400 direct sales and service representatives, 1,100 business partners, and 200 independent representatives serve our customers in over 140 countries around the globe. For more information, visit www.zoll.com.

ADVANCING RESUSCITATION. TODAY.®

ZOLL Medical Corporation
Worldwide Headquarters
269 Mill Road
Chelmsford, MA 01824
978-421-9655
800-348-9011
www.zoll.com

For subsidiary addresses and fax numbers, as well as other global locations, please go to www.zoll.com/contacts.

©2011 ZOLL Medical Corporation. All rights reserved. Advancing Resuscitation. Today., AED Plus, AED Pro, Real CPR Help, See-Thru CPR, and ZOLL are trademarks or registered trademarks of ZOLL Medical Corporation in the United States and/or other countries. All other trademarks are the property of their respective owners.

AED Plus Pelican Case

Total Protection for the First and Only Full-Rescue AED

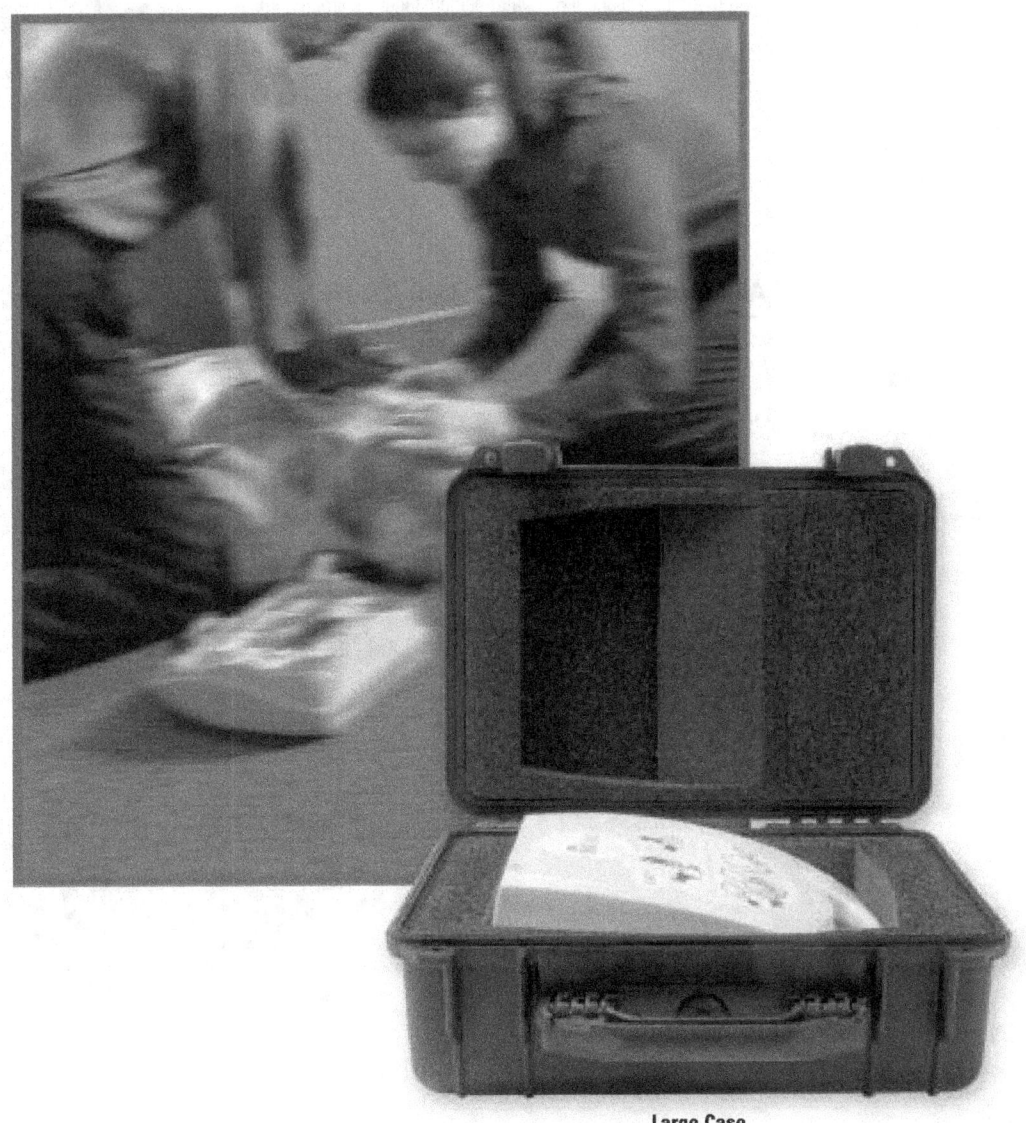

Large Case
(19" x 15.4" x 7.6")
(48.3 cm x 39.1 cm x 19.3 cm)

Unbreakable, Watertight, Dustproof, Chemical Resistant, and Corrosion Proof.

AED Plus Pelican Case

- Waterproof
- Crushproof
- Dustproof
- Corrosion proof
- Chemical resistant

Total Protection for the First and Only Full-Rescue AED

Providing an excellent shock-resistant carrier for your ZOLL® AED Plus®, the Pelican Case and its molded high-density foam provide total protection from the rigors of impact, vibration, and shock. Made of high-impact structural copolymer, it is extremely strong and durable. With its exclusive 0.25" (6.4 mm) neoprene O-ring, and ABS latches, this case seals perfectly. The Pelican Case also includes an automatic purge valve for quick equalization after changes in atmospheric pressure. The AED Plus Pelican Case is NATO-codified and tested to MIL C-4150J (Military Standard) and IP67 (water and dust ingress protection).

You break it, we replace it... forever.

A Rugged Case for the First and Only Full-Rescue AED

AED Plus Pelican Case Specifications

*ZOLL offers two cases to meet your needs: a large case for easy storage of **CPR-D•padz**™ and **pedi•padz**®**II** or a smaller case for the AED Plus alone. Each case comes complete with custom-molded foam inserts for a perfect fit.*

Large Case (8000-0837-01)
Large Case with cutouts for AED Plus, **CPR-D•padz**, or **pedi•padz II**

Inside Dimensions: 17.9" x 12.8" x 6.8" (45.5 cm x 31.8 cm x 17.3 cm)
Outside Dimensions: 19" x 15.4" x 7.6" (48.3 cm x 39.1 cm x 19.3 cm)
Temperature Ratings: Min: -10° F (-23° C) Max: 210° F (99° C)
Buoyancy: Floats in salt water with 40 lbs. (18.1 kg) load
Color: Black
Approvals: IP67 Military Standard C-4150J

Small (8000-0836-01)
Small Case with cutouts for AED Plus only

Inside Dimensions: 14.7" x 10.4" x 6" (37.3 cm x 26.4 cm x 15.2 cm)
Outside Dimensions: 16" x 13" x 6.9" (40.6 cm x 33 cm x 17.5 cm)
Temperature Ratings: Min: -10° F (-23° C) Max: 210° F (99° C)
Buoyancy: Floats in salt water with 20 lbs. (9.07 kg) load
Color: Black
Approvals: IP67 Military Standard C-4150J

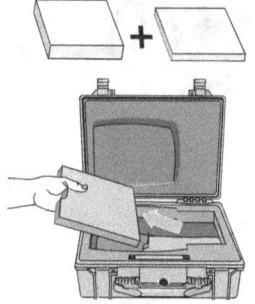
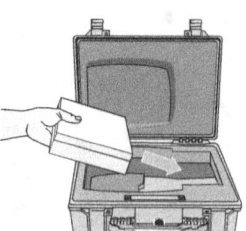
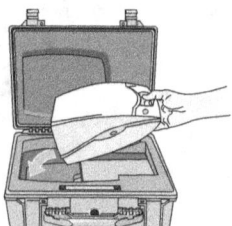

ZOLL's AED Plus Large Pelican Case allows you to store your AED Plus and an extra set of **CPR-D•padz** with the option of also storing a set of **pedi•padz II**.

ZOLL Medical Corporation Worldwide Headquarters
269 Mill Road
Chelmsford, MA 01824
978-421-9655
800-348-9011

ZOLL Direct Sales Subsidiaries
Australia
+61 2 8424 8700
www.zoll.com.au

Austria
+43 650 413 6222
www.zollmedical.at

Canada
905 629 5005
www.zoll.com

France
+33 (1) 30 05 14 98
www.zoll.fr

Germany
+49 2236 87 87 0
www.zollmedical.de

The Netherlands
+31 (0) 488 41 11 83
www.zoll.nl

United Kingdom
+44 (0) 1925 846 400
www.zoll.com

For subsidiary addresses and fax numbers, as well as other global locations, please go to www.zoll.com/contacts.

Specifications subject to change without notice.
©2007 ZOLL Medical Corporation. All rights reserved. "Advancing Resuscitation. Today." and **CPR-D•padz** are trademarks of ZOLL Medical Corporation. AED Plus, **pedi•padz II** and ZOLL are registered trademarks of ZOLL Medical Corporation.
Printed in U.S.A. 040710 9656-0173

Ingress Protection Case Study #1
Ottawa Police Service and the Submerged AED Plus

Ottawa Police Service

The Story

Constable Rick Giroux of the Ottawa Police Service is assigned to the harbor patrol. His boat is outfitted with a ZOLL AED Plus™ automated external defibrillator (AED). He recalls a rather unique situation back in October that involved this AED.

"Our patrol boat was docked because of an impending hurricane," said Giroux. "We had a pump going inside the boat to prevent it from sinking. Once the storm hit, however, the pump failed, and the boat took on water."

As the boat filled with water, it flipped over and ripped the cleats right off the pier. Although the boat didn't sink, all the equipment, including the AED Plus, was submerged for four hours.

After the hurricane, Giroux said that a Marine Patrol Officer removed the waterlogged AED, attempted to turn it on, and found that it worked fine. Giroux then brought it to the Biomedical Department of the Ottawa Paramedics Service, which took it out of service, despite the fact that it was still functioning. They sent the soaked AED to ZOLL Medical Corporation's Technical Service Department for inspection.

ZOLL's technical team performed extensive environmental and operational testing on the AED Plus, which proved the AED was fully functional. Without hesitation, ZOLL TechnicalService recertified the AED Plus for use in the field.

The ZOLL AED Plus has been tested for particle and water ingress, and has received a rating of IP55. It has the highest rating of any AED available today. At ZOLL, our claims for meeting these test specifications are conservative. The experience of the Ottawa Police Service shows that in some cases, the AED Plus can survive hours of complete submersion without any loss of functionality.

Ingress Protection

Definition

A worldwide standard has been established by the International Electrotechnical Commission (IEC) for comparing the ability of electronic devices to withstand exposure to dust particles and water. The IEC describes its mission to be:

> ...*the leading global organization that prepares and publishes international standards for all electrical, electronic and related technologies. These serve as a basis for national standardization and as references when drafting international tenders and contracts. Through its members, the IEC promotes international cooperation on all questions of electrotechnical standardization and related matters, such as the assessment of conformity to standards, in the fields of electricity, electronics and related technologies.* [Quoted from the IEC website.]

The ratings established by the IEC for resistance to particulates and water are called "Ingress Protection" or "IP" ratings, and are defined by IEC 60529, Degrees of protection provided by enclosures ("IP" Code), for all IP Codes. A full rating contains two digits, each of which can take a value of 1 through 6. The first measures the ability of the device to resist the ingress of foreign objects, or dust. The second measures the ability to resist the ingress of moisture. The first digit can have a value from 1 to 6, the second a value of 1 to 8. The higher the number, the better the protection. The lowest combined rating would be IP11; the highest would be IP68. Where a device has not been rated for either dust or water, an "X" is substituted for the digit. Thus a device, like Medronic's CR Plus, with a rating of IPX4, has not been tested and rated for its ability to resist dust, while its rating for water ingress is 4. So what do the different digit values mean? The table below was developed by Underwriter's Laboratory (UL) to explain the Ingress Protection code values, as specified in the IEC 60529 standard.

IP Ratings Explained by Underwriters Laboratories (UL)

First Digit	Protection Against Foreign Objects	Second Digit	Protection Against Moisture
0	Not protected	0	Not protected
1	Protected against objects greater than 50mm	1	Protected against dripping water
2	Protected against objects greater than 12mm	2	Protected against dripping water when tilted up to 15°N
3	Protected against objects greater than 2.5mm	3	Protected against spraying water
4	Protected against objects greater than 1.0mm	4	Protected against splashing water
5	Dust Protected	5	Protected against water jets
6	Dust Tight	6	Protected against heavy seas
		7	Protected against the effects of immersion
		8	Complete protection against submersion

Ingress Protection Comparison

ZOLL AED Plus	Medtronic LIFEPAK CR Plus	Medtronic LIFEPAK 500	Philips HeartStart OnSite	Philips HeartStart FR2+	Cardiac Science Powerheart G3	Welch Allyn AED10
IP55	IPX4	IPX4	IP21	IP54	IP24	IP24

Particle Size Comparison

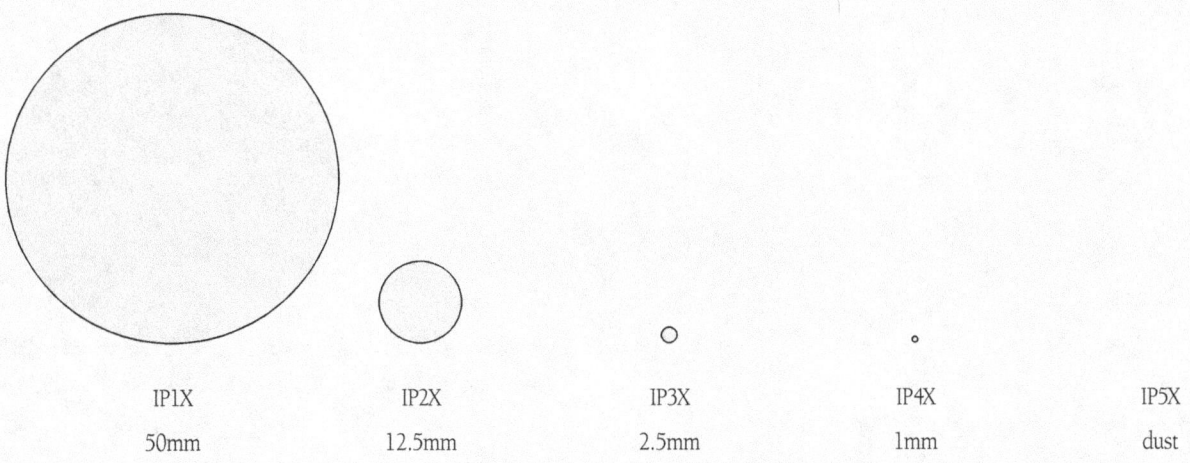

| IP1X | IP2X | IP3X | IP4X | IP5X |
| 50mm | 12.5mm | 2.5mm | 1mm | dust |

To get some notion of how big the "foreign objects" being described in each of these ratings really are, the diagram above shows actual particle sizes. This shows that an AED like the Philips HeartStart OnSite, because it has a rating of IP21, is in danger of being compromised by a small pebble about 1/2 inch in diameter.

Only one other public access AED has as an IP rating for particulate matter at five: the Philips FR2+.

All other public access AEDs can be incapacitated by any foreign object smaller than the little 1 mm dot above. If there is any dust or spraying water present in your environment, and you don't have an AED Plus, you're running a risk that when you need your AED, it may not work. The table above lists the IP ratings of all of ZOLL's competitors.

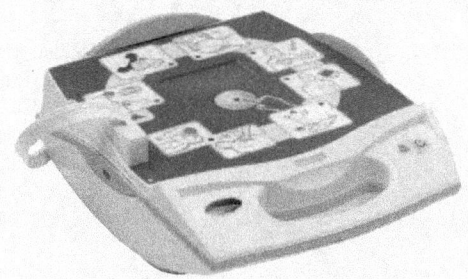

ZOLL Medical Corporation
Worldwide Headquarters
269 Mill Road
Chelmsford, MA 01824
USA
(978) 421-9655
www.zoll.com

©2005 ZOLL Medical Corporation. All rights reserved. ZOLL and AED Plus are trademarks of ZOLL Medical Corporation. All trademarks are property of their respective owners.

Existing Underground Utility Locations (NYC DDC)

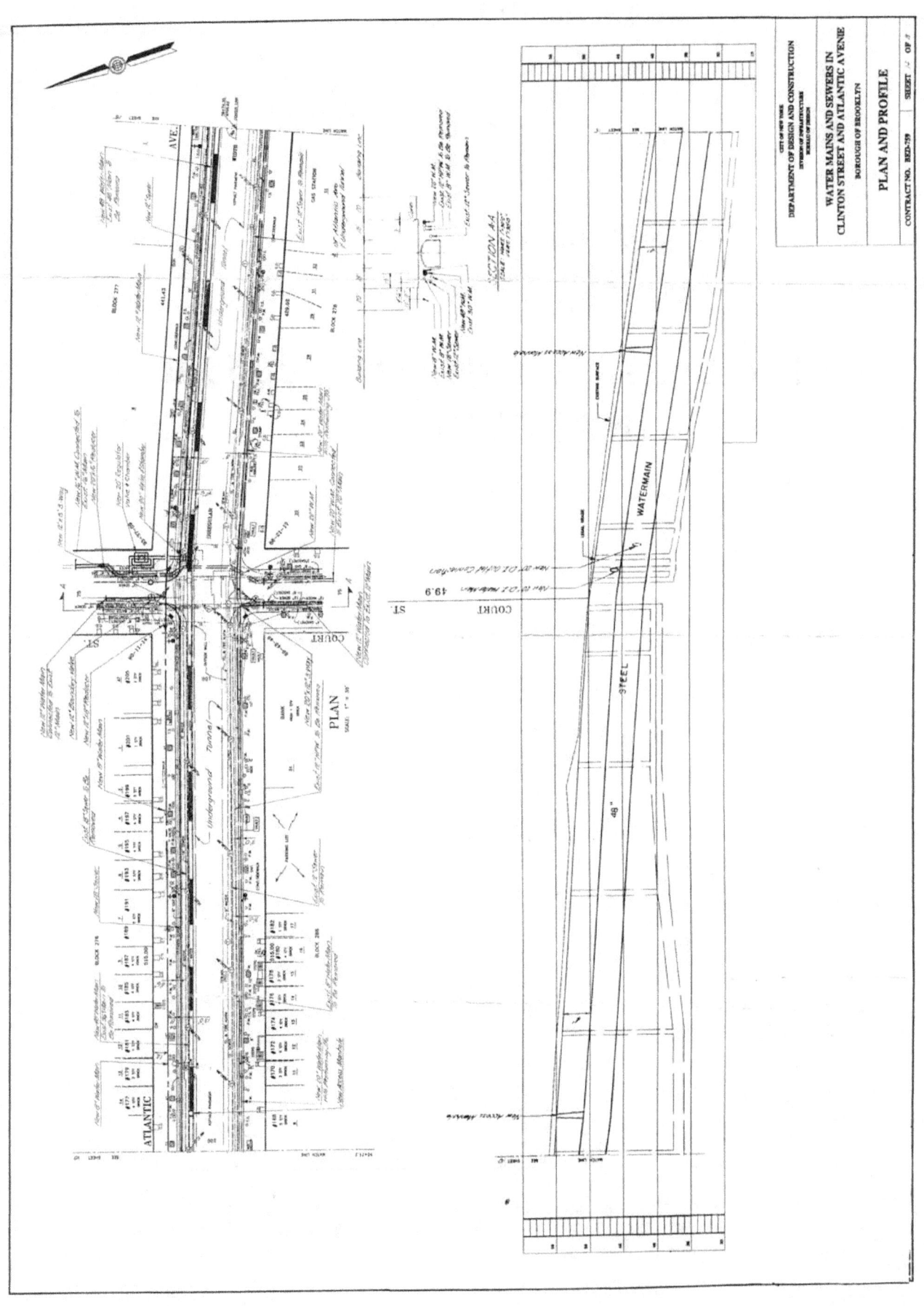

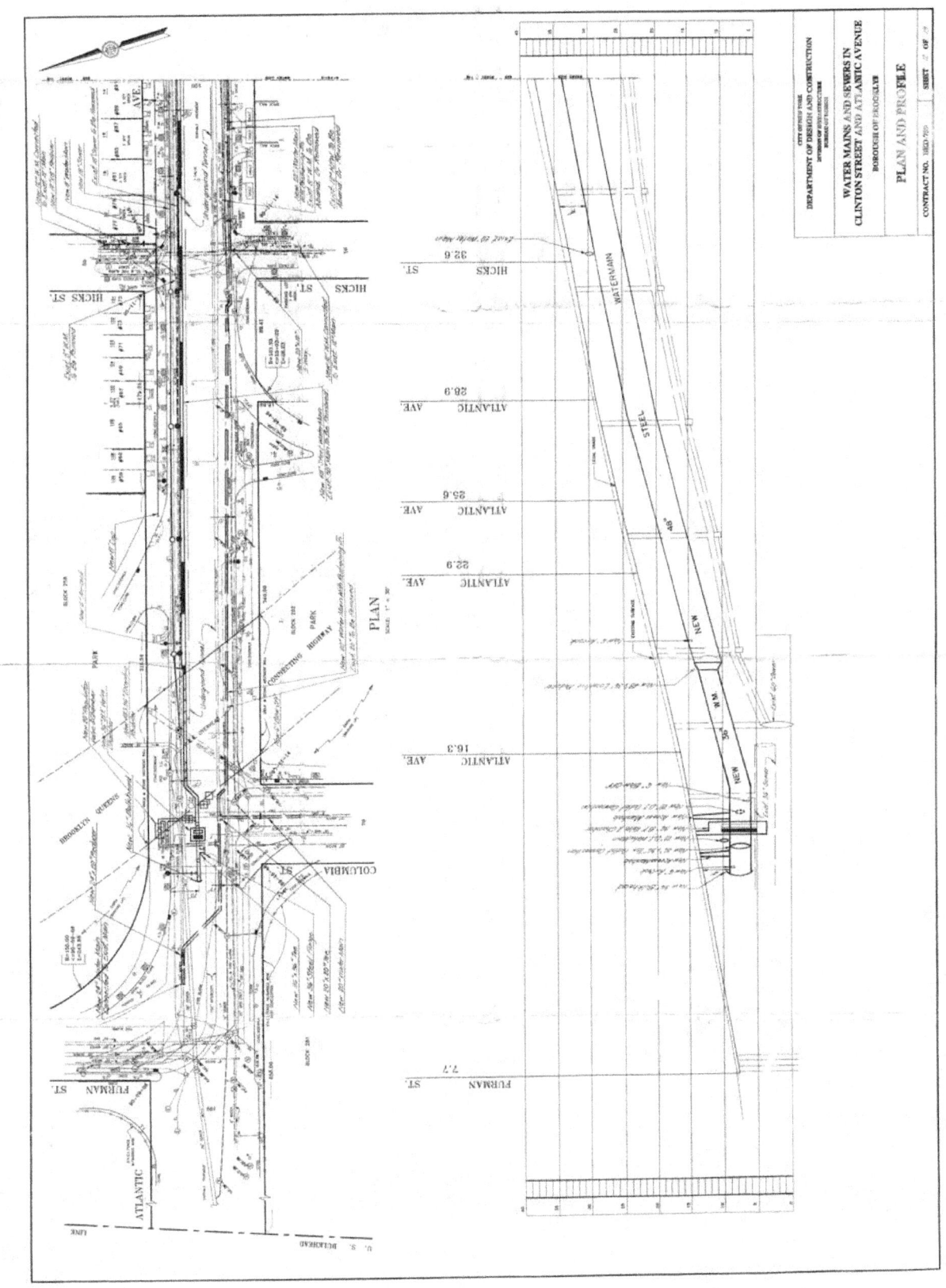

Elevator Cost - MTA Paid $ 7 Million

MTA Paid $7 Million For Elevator

SPECIAL SERIES
the americans with disability act at 25

In Helping Those With Disabilities, ADA Improves Access For All

JULY 24, 2015 4:19 AM ET

JOSEPH SHAPIRO

Listen to the Story
Morning Edition
7:05

- Playlist
- Download
- Embed
- Transcript

When the Americans with Disabilities Act (ADA) became law 25 years ago, "everybody was thinking about the iconic person in a wheelchair," says civil rights lawyer Sid Wolinsky. Or that the ADA — which bans discrimination based on disability — was for someone who is deaf, or blind.

President George H.W. Bush signed the Americans with Disabilities Act on the South Lawn of the White House on July 26, 1990. "Let the shameful wall of exclusion finally come tumbling down," he said that day.
Barry Thumma/AP

But take a tour of New York City with Wolinsky — and the places he sued there — and you will see how the ADA has helped not just people with those significant disabilities, but also people with minor disabilities, and people with no disability at all.

Stop 1: Dyckman Street Subway Station, Inwood, Manhattan
At this subway station at the northern tip of Manhattan, Wolinksy — who is joined by Jim Weisman, an attorney with the United Spinal Association — points to the elevator that was added after they sued the city in November 2013. Now, wheelchair users can get to the platform.

Attorneys Sid Wolinsky (left) and Jim Weisman specialize in accessibility lawsuits to ensure the Americans with Disabilities Act is being properly enforced.
Adam Wolffbrandt/NPR

On this day, the elevator is in constant use, gliding up and down, the doors opening with a ding as people use it instead of climbing the steep stairs nearby.

But Dustin Jones, a wheelchair user who joins us, notes: "I have not seen a person with a disability yet ride that elevator. It's all been walking people."

Over the course of an hour, no one — other than Jones — is in a wheelchair. Jones watches a mother get on, holding the hand of one young child and pushing a baby in a stroller, while carrying bags.

"This is one of those stations where it would be really tricky to navigate a small child, a small baby with the stroller and bags, if you had to solely use the steps," he says.

Weisman, a veteran of accessibility lawsuits, also doesn't see any people with any visible disabilities enter the elevator.

"See, the elevator use is constant. So there must be a reason: Elderly people, people with vertigo and balance problems and knee problems and coordination, people choose to use the elevator," he says.

"This elevator is a gift from the disability community and the ADA to the nondisabled people of New York," says Wolinsky, who co-founded Disability Rights Advocates.

The elevator at the Dyckman Street Subway Station in Inwood, Manhattan, helps people of all abilities reach the platform.
Michael Rubenstein for NPR

The ADA, which was signed into law on July 26, 1990, banned discrimination based on disability in employment. It also required access to government services and public accommodations, and included a rule that public transportation be made accessible.

That means adding an elevator to a major subway station, or including elevators when building new stations, or adding one when there's a substantial renovation at a station.

So when the Dyckman Street Station was renovated, the lawyers sued to have an elevator included. It can sometimes be expensive to provide access. In this case, the tricky work of installing the elevator at an old station located on a rocky cliff cost about $7 million, or a little more than 20 percent of the overall cost of the renovation. And that was just for one elevator, to the downtown trains only.

NPR analyzed transit systems in New York and 16 other cities across the country. In New York, only 21 percent of subway stations are accessible. That's the highest percentage of inaccessible stations of any system in the country. New York's subway has the nation's highest number of riders, but most of the city's current stations were built before 1940, long before wheelchair accessibility became an issue.

Kevin Ortiz, a spokesman for New York's Metropolitan Transportation Authority, notes that the system has made progress, and has committed hundreds of millions of dollars on a promise to make 121 stations wheelchair accessible by 2020. That would bring up the percentage of accessible stations in the city to roughly 25 percent. In addition, Ortiz says, "every single one of our 5,700 buses is accessible," another change mandated by the ADA.

The NPR analysis found that older transit systems remain the least accessible. After New York's subway system, the system with the lowest number of accessible stations is the Philadelphia metro area's commuter rail system, at 43 percent.

In other systems, newer subway stations, including any built after the ADA, are accessible. The stations in the transit systems in the Washington, D.C., metro area, Los Angeles, Dallas, Portland, Oregon and Denver are 100 percent accessible, according to the NPR numbers.

Stop 2: Lillian Wald Houses, Lower East Side, Manhattan
"When Hurricane Sandy hit, I was here, in my apartment," says Melba Torres, 53, who lives at this red-brick public housing complex on Manhattan's Lower East Side. "I live on an eighth floor, and I was not able to be evacuated."

When the October 2012 storm knocked out power to her building, the elevator shut down. She has cerebral palsy, and her power wheelchair weighs a few hundred pounds. Like thousands of people in high rises around the city, she was stuck.

Torres stayed in her apartment for six days without power.

"It's the fear of not knowing what could happen. It was totally dark. I don't like darkness," she says.

When the power went out during Superstorm Sandy, Melba Torres was trapped in her eighth-floor apartment on the Lower East Side of Manhattan for several days without access to a working elevator.
Adam Wolffbrandt/NPR

In the daylight, as the storm surged, she looked out her window and watched the water rising in the nearby East River.

"I was so scared. I had never seen anything like that. And I had never seen water come out of the East River that way. It was like being in a movie," she recalls.

For a person with a disability, losing electricity isn't just an inconvenience — it can be a matter of life and death. Some people with disabilities lost power to home dialysis machines, or even to portable respirators that helped them breathe.

"Everything that I use to live on: my chair, my device to get in and out of bed," she says. "My bed is electric."

Torres also depends on her elderly mother, who lives down the block. But her mother couldn't walk up the eight flights of stairs to help. "My mom was very frightened for me," says Torres. "She's in her 70s and she has her own health issues, and she refused to go to a shelter knowing that I was here."

Related NPR Stories

ord," he says.

Jennifer Carione puts food on Dan Carione's plate while their daughter Sophia eats dinner at their home in Brooklyn, N.Y.
Michael Rubenstein for NPR

Carione sued, using the ADA. He argued that, with his hearing aid, he could do his job as well as or better than before. In March, the city settled, and recently, Carione, who is 48, went back to work. Now New York will set new rules for letting police use hearing aids.

"The ADA offers protection for everybody," says Wolinsky, who sued on behalf of Carione. "In fact, this is an enormous group of Americans — people who don't define themselves as disabled. The person who's in their 80s and moving really slowly, and can't manage a flight of steps, doesn't think of themselves as disabled — they're just a little older. A person with a mild case of arthritis, a person who can't manage a heavy suitcase when they're traveling — those are the people who are being helped by the ADA, and it's a large and growing population."

That's clear from U.S. Census Bureau numbers. About 1 in 5 Americans has a disability. And the number keeps going up as people age: About 40 percent of people 65 or older report they have one or more disabilities.

Examples of "Tunnel Experience" Content

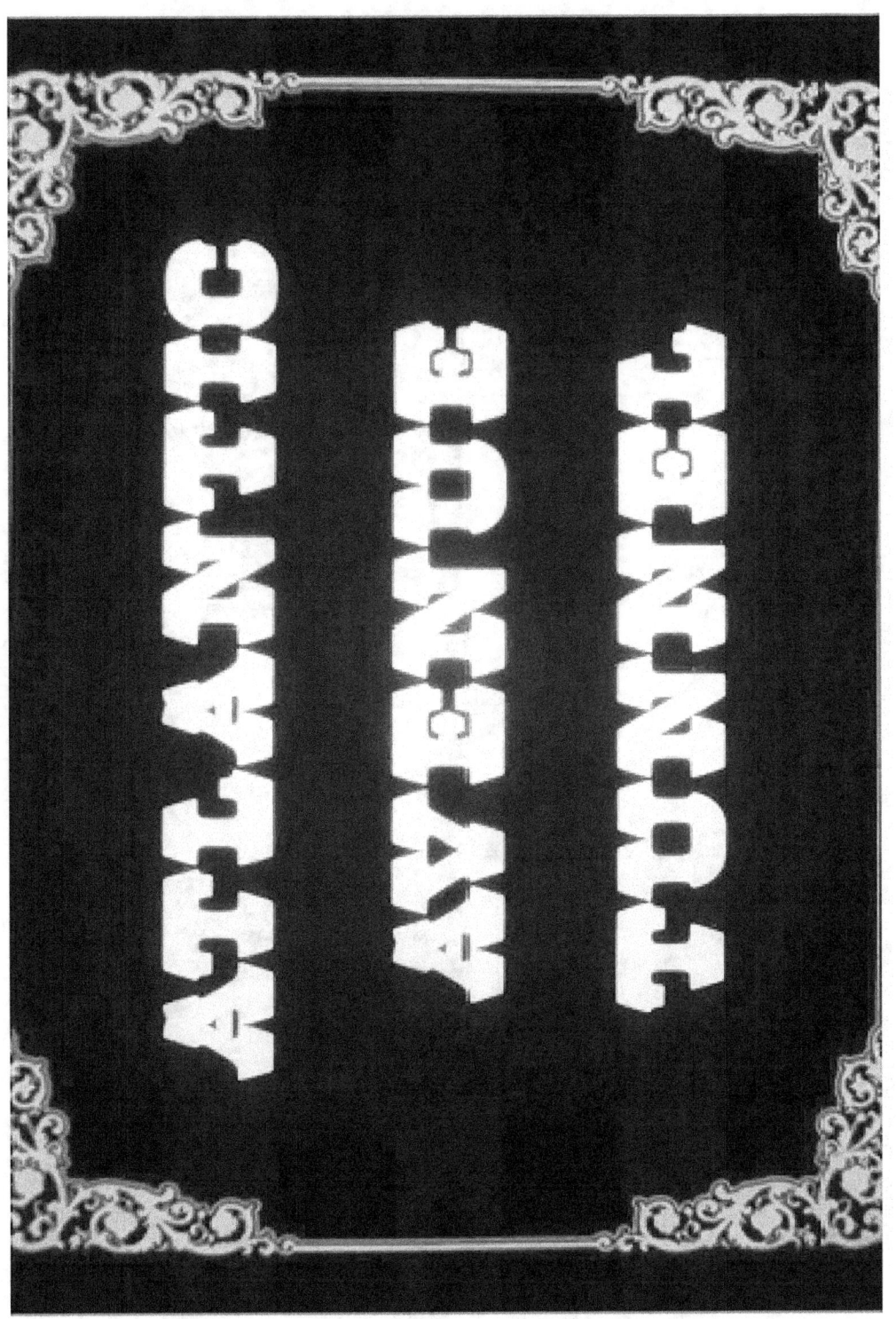

By Bob Diamond, © 1979- 2015

Cunard *BRITANNIA* [11] in the ice at Boston, February 1844

(*Mariners' Museum*)

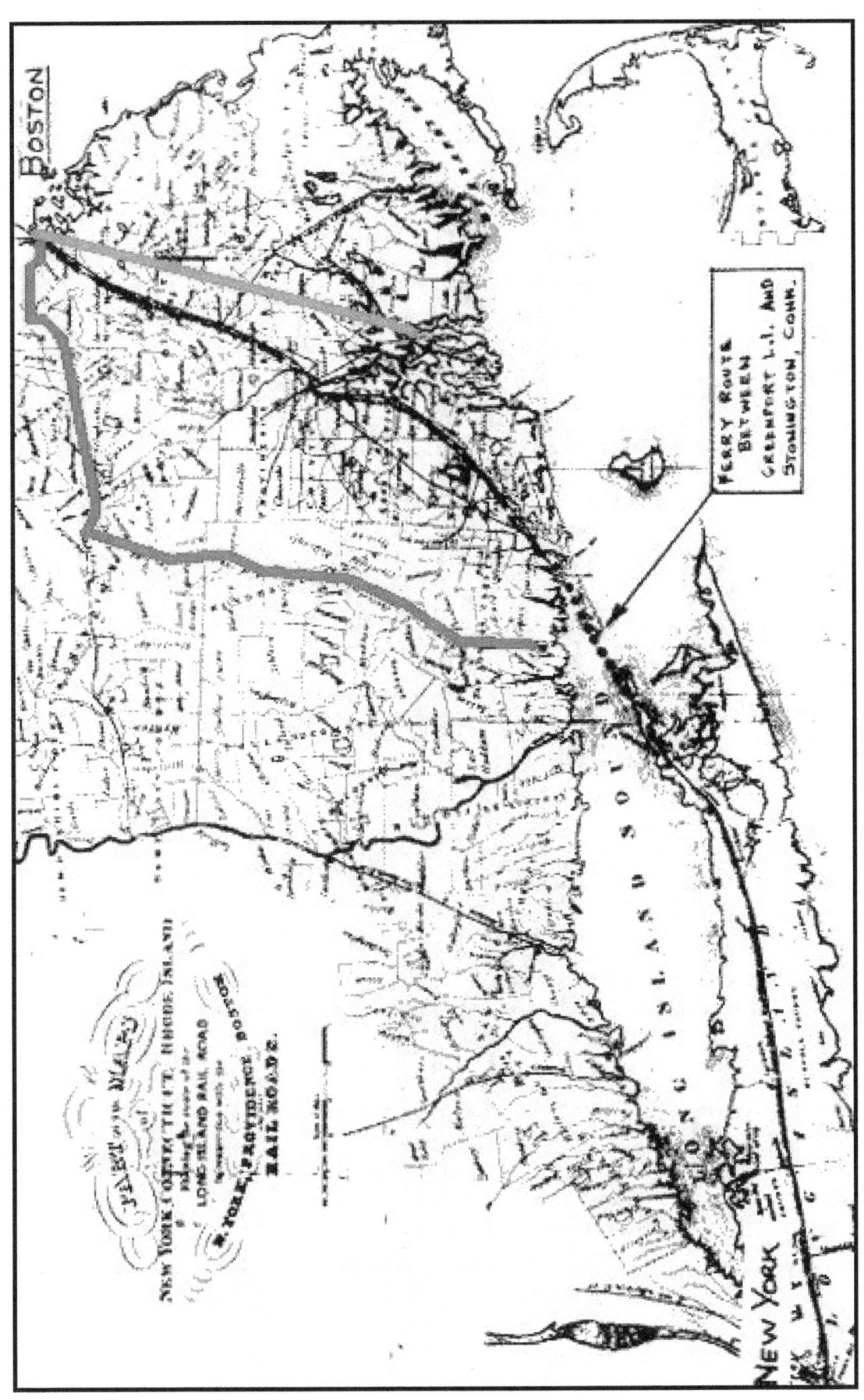

Original map showing Long Island Railroad as part of New York to Boston Rail System (circa 1844)

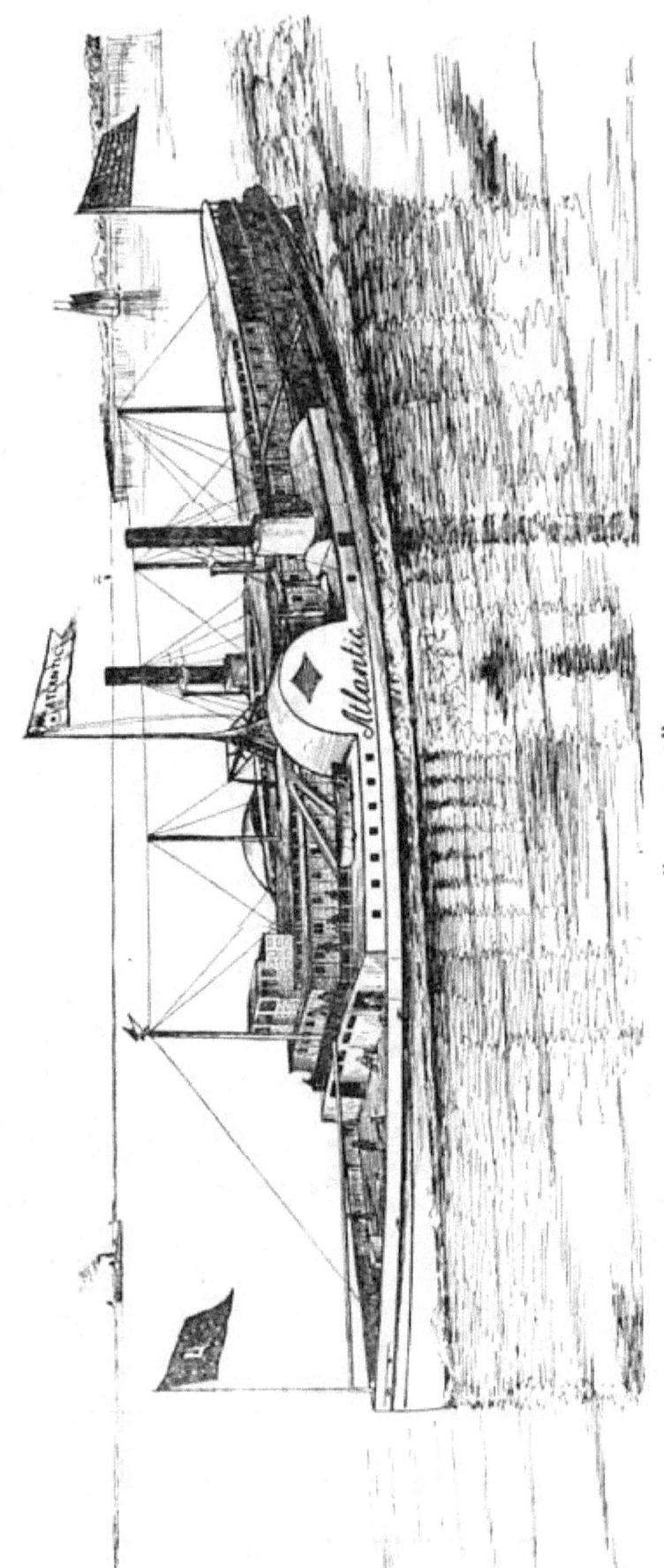

"ATLANTIC."

Illustration from Harper's Weekly September 23, 1865, depicting the hazards of steam powered travel. These hazards include: boiler explosions, fires, derailments, bridge collapses, etc. Even steamboats (such as the Lexington) were sometimes destroyed by fire.

The use of the streets by the locomotives of the Long Island Rail Road posed a problem to the municipality. Trains traversed Atlantic Avenue after 1836 so as to reach the terminal which was located in the heart of the city. At first, the Company was permitted to use only horses in drawing the cars through the streets of the city. Soon after the line began operating, however, the authorities decided to allow steam engines to enter the city if their speed was confined to six miles per hour.[33] This arrangement appeared to be satisfactory until accidents occurred in 1839 in which two youths were killed. The matter was then brought before the Common Council for renewed consideration.[34] The aldermen determined to place more restrictions on the use of steam locomotives in the center of the community. The speed limit was again limited to six miles an hour and it was further stipulated that the use of locomotives would be prohibited between the hours of sunset and sunrise. Also, the engine was to be equipped with an attachment "calculated to take up any object or person lying or being upon the said railroad."[35] The train had to sound a warning when it approached a crossing and ring a bell when traveling between Clinton and Atlantic Avenues.[36]

Common Council.

January 15th, 1844.

The Board met pursuant to adjournment.

Present—His Honor the Mayor, presiding; and Aldermen Van Wagenen, Thorne, West, Cornell, Greene, Low, Brigham, Gerald, Smith, Boulton, Osborne, Bergen, Boerum, Kelsey.

The minutes of the last meeting were read and approved.

The several petitions of C. Davis and others, and of William Cook and 170 others, requiring the Long Island Rail Road Co. to remove the cars and engines from the street to the wharf and ground near the foot of Atlantic st., and cut through, or tunnel through, the hill on said st.; were referred to the Street Com.

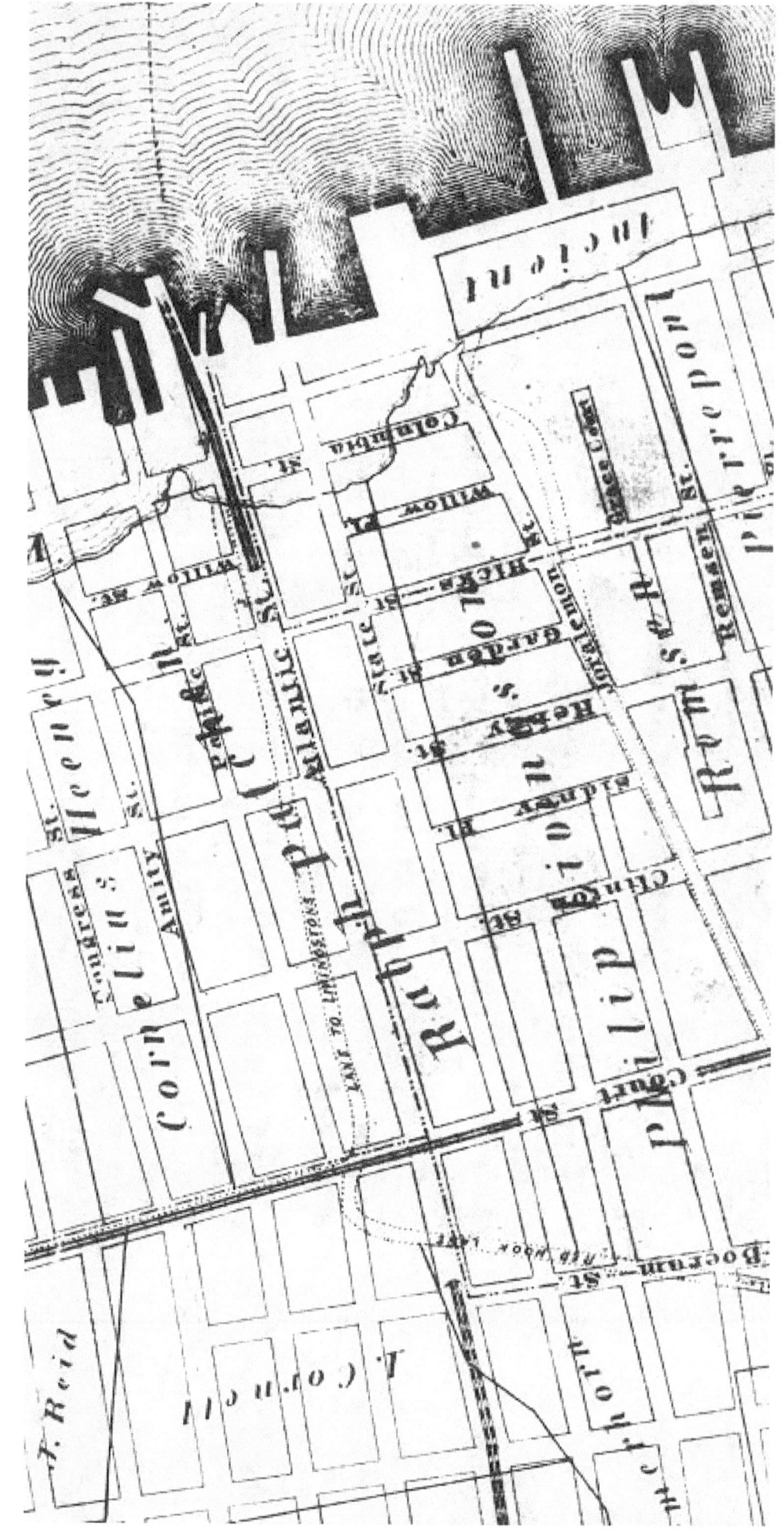

- **"Commodore" Cornelius Vanderbilt** was the Operations Director and a Board Member of the LIRR at the time the tunnel was built. The LIRR was Vanderbilt's first railroad enterprise- not the New York Central RR, as is the popular thought.

Sketch depicting the cut and cover technique in the mid 19th century.

Source: A Treatise on Explosive Compounds, Machine Rock Drills and Blasting, 1878

Tunnel Entrance, 1844

Tunnel Composite Map - Historic Maps (1846, 1851, 1855, 1856 and 1886) Overlayed Squared over Current 2009 Tax Map

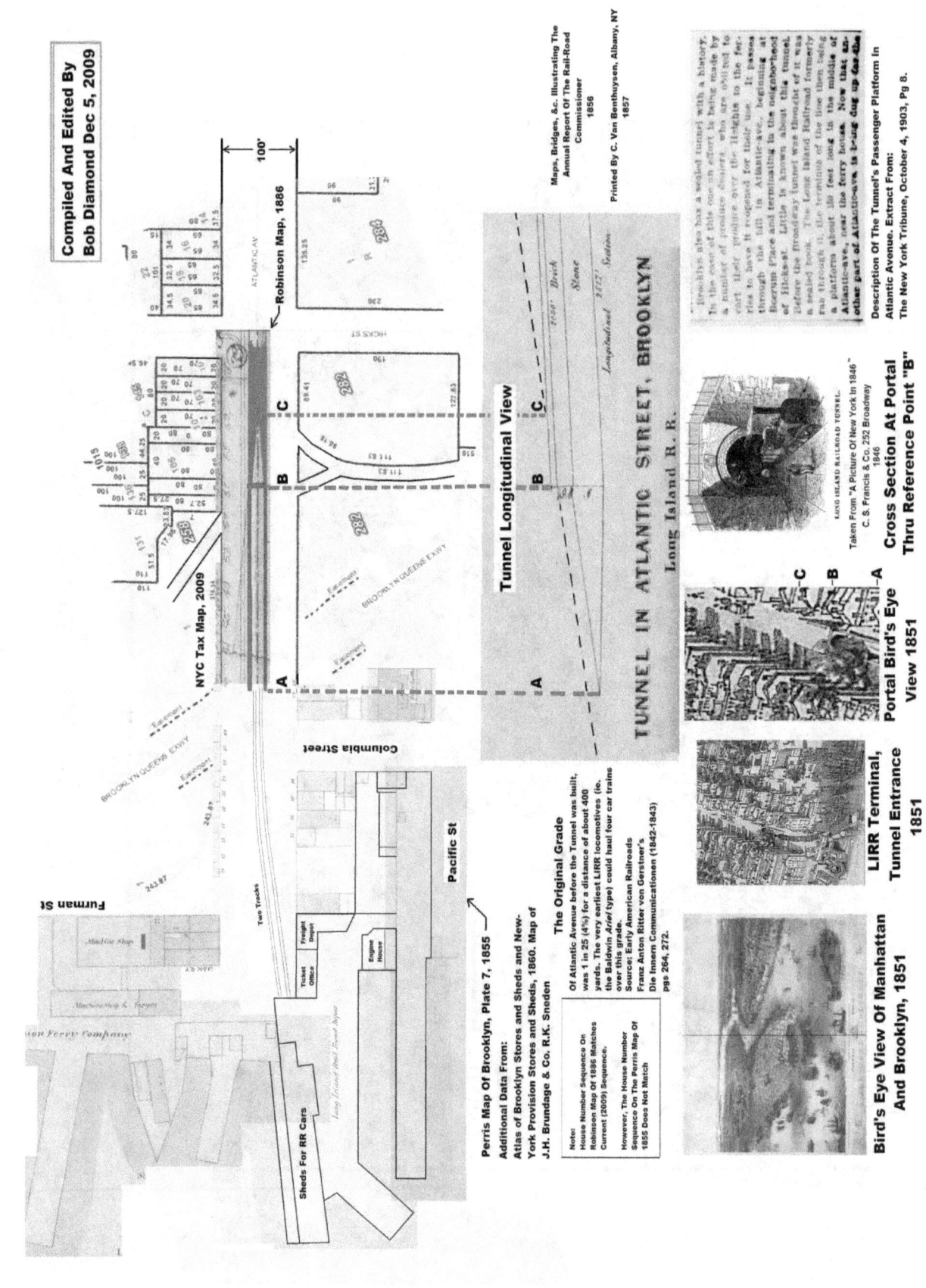

E.B. Litchfield

Sketch of proposed "Atlantic Avenue, Drive and Promenade" project, A massive real estate development plan.

ATLANTIC AVENUE COMMISSION
CITY OF BROOKLYN
1897
THE COST OF THE REMOVAL.

This act to provide for the closing of the tunnel and removing steam from Atlantic avenue was amended by Chapter 100 of the Laws of 1860; but the change related mainly to the method of the collection of the assessments to be laid on the adjoining property, and is not particularly important in this connection.

It is not necessary to reproduce the proceedings in full which were taken under the acts for the closing of the tunnel and the removal of steam from Atlantic avenue.

It is enough to say that in accordance with the act the common council of the city of Brooklyn, on the first day of August, 1859, presented a petition to the Supreme Court, asking for the appointment of commissioners as provided by the act, and that an order was made on that day naming such commissioners; that the commissioners made their report on the 28th day of July, 1860, determining what amount should be paid to the Long Island Railroad Company as compensation, fixing such amount at the sum of $125,000, which, together with the expenses of the commission, amounting in the aggregate to the sum of $129,801.80, was assessed upon the property lying along the avenue; that on the 27th day of September, 1861, the court made an order duly confirming the report of the commissioners, and that thereafter this amount was collected from the owners of the adjoining property; the amount so collected was paid to the Brooklyn and Jamaica Railroad Company, which had then become the assignee of the Long Island Railroad Company in respect to that fund; and that the tunnel was thereafter closed and steam was removed from Atlantic avenue as provided by the act in question.

No. 28

PLEASE BRING THIS BILL WITH YOU.

Office Brooklyn Central and Jamaica Rail Road Company.

No. 1 ATLANTIC STREET, BROOKLYN.

The Commissioners appointed under the act of April, 1859, for closing the Atlantic Street Tunnel, were authorized and directed by the amended Act of March 23d, 1860, to make and record their Assessment List with the Register of the County of Kings, and to make and deliver to the Long Island Rail Road Company, or its assigns, an assignment of said list, and upon receiving the assignment, the Long Island Rail Road Company or its assigns, are authorized to receive for their own use, the amount which any person shall be liable to pay and upon such payment being made, the property of the person making the same shall be released from the said assessment.

As authorized under the Acts referred to above, the Commissioners have executed an assignment to the Brooklyn and Jamaica Railroad Company, (assigns of the Long Island Railroad Company,) of their Assessment List and said Company are fully authorized to collect the same under said Acts.

(Signed,)

THEODORE F. KING,
JOHN D. LAWRENCE,
JOHN WINSLOW,
Commissioners for Closing Tunnel, Atlantic Street.

CONFIRMED, September 27th, 1861.

Mr. Henry D. Young

To BROOKLYN AND JAMAICA RAIL ROAD COMPANY, Dr.

N. B.—If unpaid at the expiration of thirty days from date of the confirmation, two per cent. will be added, at the expiration of sixty days, three per cent. for Collector's fees. And interest at the rate of one per cent. per month until paid.

Assessment Number.	Frontage	AT STREET.	BETWEEN WHAT STREETS.	DOLLARS	CTS.
3	20	Atlantic	Furman & the River		

LONG ISLAND HISTORICAL SOCIETY
128 PIERREPONT STREET
BROOKLYN, N.Y. 11201

JUN 13 1984

COLLECTOR'S FEES, 3.90
1% 9 mo. INTEREST, 9.10
$143.00

Received Payment, this 22 day of May 1862

Theo. Warren, Collector.

http://www.landgrant.org/history.html

1857 Minnesota & Northwest rechartered Minnesota & Pacific. Granted five million acres and several million dollars in Minnesota state bonds. Built only 10 miles of railroad. Insolvency led to foreclosure in 1860. Reorganized into the St. Paul & Pacific and the First Division, escaping debts but not relinquishing grants or franchise rights. Increased grant to ten sections per mile. Mortgaged railroad and grants to Dutch capitalists for $13 million. Some $8 million was siphoned to phony construction. Renamed the Great Northern in 1885-1889.

Slow Train to Paradise: How Dutch Investment Helped Build American Railroads By A. J. Veenendaal

Not all was lost, however, because the same promoters incorporated a new company in the same year and named it the St. Paul & Pacific Railroad Company, which again planned to build a main line from St. Paul west to Breckenridge with a branch to St. Vincent on the Canadian border. The land grant of its predecessor was also taken over.[2] Construction remained slow until a contract was signed with Electus B. Litchfield and Company of Brooklyn, New York, for construction of part of the line in exchange for shares of the railroad plus some cash.[3] By that time the first ten miles, between St. Paul and what is now Minneapolis, were in operation.

A first loan of $120,000 had already been issued, secured by a first mortgage on those ten miles, and in 1862 a new 7-percent issue of $700,000 (per 1892) was floated, secured by a second mortgage on the same ten miles, plus a first mortgage on 70 miles of the branch line. As this was not yet enough to cover construction costs, a third issue was floated the same year, again at 7 percent, but this time for $1.2 million and secured by a third mortgage on the first 10 miles, a second mortgage on the 70 miles of the branch, plus a first lien on 307,200 acres of the land grant.

Annual Report By Minnesota. Office of Railroad Commissioner, 1873

FUNDED DEBT

Consists of first mortgage bonds on 150 miles of road	$1,500,000 00
Second mortgage bonds on 150 miles of road, and first on lands	3,000,000 00
Third mortgage bonds on 150 miles of road and lands, and first on additional lands and road	6,000,000 00
Fourth mortgage bonds on lands and road	3,000,000 00
Total funded debt	$13,500,000 00
Number of stockholders	Unknown.
Stock held in Minnesota—little, not to exceed	$50,000 00
Total amount of dividends paid	None.

NOTE: In Todays Money, That $13.5 Million is Worth $40.5 BILLION !

Source: http://www.measuringworth.com/uscompare/

CHARACTERISTICS OF ROAD.

The length of the road, single main track, is 76 miles, laid with "T" rail, (both English and American,) weighing from 45 to 58 lbs. per yard.

The aggregate length of sidings and other tracks not above enumerated, is about	6⅜ miles.
The number of switches on the road, (none of which are private) is	64
Number of crossings	40
(None of the crossings are provided with flagmen.)	
Number of crossings level	38
Number of crossings bridged	2
Number of stations on road	11
Number of engine houses and shops	4
Number of water stations	7

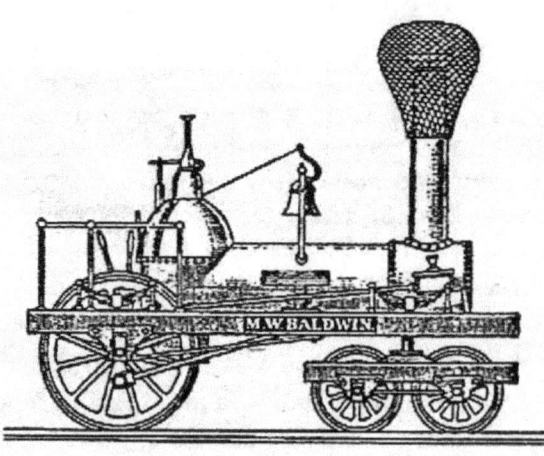

"A PASSAGE OF SOLEMNITY AND DARKNESS"
By Walt Whitman

Editor's note: In the summer of 1861, Walt Whitman, already a celebrity in New York since the publication of "Leaves of Grass," wrote a series of newspaper columns for the Brooklyn Standard under the rubric "Brooklyniana." The following excerpt is the poet's own musing on the Atlantic Avenue Tunnel.

"The old tunnel, that used to lie there under ground, a passage of Acheron-like solemnity and darkness, now all closed and filled up, and soon to be utterly forgotten, with all its reminiscences.... For it was here you started to go down the island, in summer. For years, it was confidently counted on that this spot, and the railroad of which it was the terminus, were going to prove the permanent seat of the business and wealth that belong to such enterprise. But its glory, after enduring in great splendor for a season, has now vanished — at least its old Long Island Railroad glory has. We were along there a few days since, and could not help stopping, and giving the reins for a few moments to an imagination of the period when the daily eastern train, with a long string of cars, filled with summer passengers, was about starting for Greenport, after touching at all the intermediate villages and depots. We are (our fancy will have it so), in that train of cars, ready to start. The bell rings, and winds off with that sort of a twirl or a gulp (if you can imagine a bell gulping), which expresses the last call, and no more afterwards; then off we go. Every person attached to the road jumps on from the ground or some of the various platforms, after the train starts — which (so imitative an animal is man) sets a fine example for greenhorns or careless people at some future time to fix themselves off with broken legs or perhaps mangled bodies. The orange women, the newsboys, and the limping young man with long-lived cakes, look in at the windows with an expression that says very plainly, "We'll run alongside, and risk all the danger, while you find the change." The smoke with a greasy smell comes drifting along, and you whisk into the tunnel.

"The tunnel: dark as the grave, cold, damp, and silent. How beautiful look earth and heaven again, as we emerge from the gloom! It might not be unprofitable, now and then, to send us mortals — the dissatisfied ones, at least, and that's a large proportion — into some tunnel of several days' journey. We'd perhaps grumble less, afterward, at God's handiwork." ∎

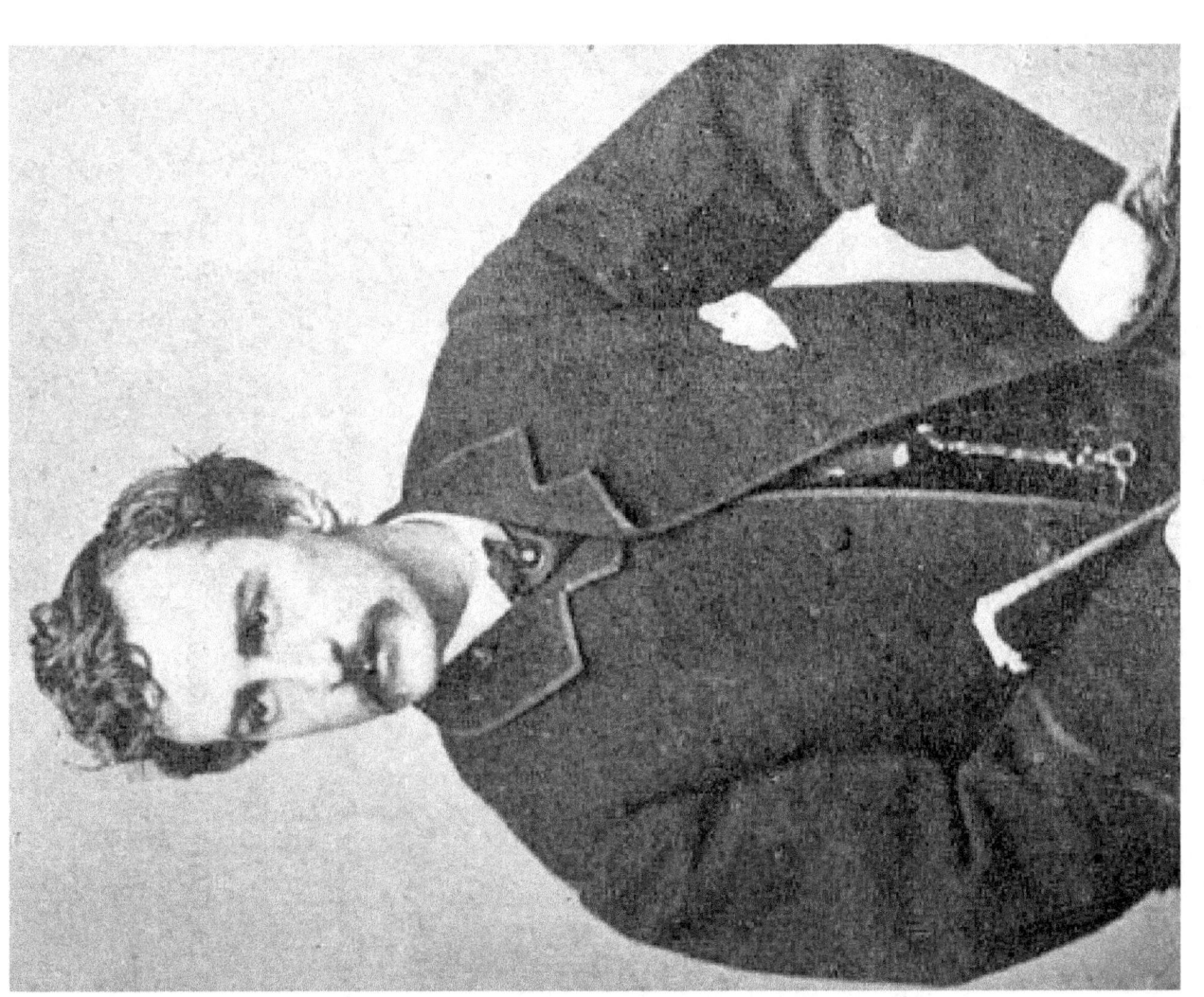

John Wilkes Booth

Did He Really Live To A Ripe Old Age, Under An Assumed Name?

ACADEMY OF MUSIC, BROOKLYN.

SATURDAY EVENING, OCT. 24TH, 1863.

First night in Brooklyn of

J. WILKES BOOTH
AND
MRS. JULIA BENNETT BARROW.

MR. BOOTH

Begs to announce that he will make his first appearance as

THE DUKE OF GLOSTER

in Shakespeare's Tragedy of

RICHARD III.

The celebrated actress

MRS. JULIA BENNETT BARROW

as

QUEEN ELIZABETH,

MISS FANNY BROWN

as

LADY ANNE.

Supported by a talented Company.

Box Book open on Friday morning.   o20 3t*

John Wilkes booth performs Shakespeare's *Richard III* at the Brooklyn Academy of Music. *Brooklyn Eagle*, October 23, 1863

BOOTH.—On Saturday, to a poor house, Wilkes Booth gave a wretchedly appointed though well acted representation of Richard the Third. To-night he plays "Raphael" in the "Marble Heart" to Mrs. Burrow's "Marco." We trust, for the honor of his name, he may have a better attendance than before.

Brooklyn Daily Eagle October 26, 1863

Chattanooga Campaign

From Wikipedia, the free encyclopedia

Chattanooga Campaign
Part of the American Civil War

Maj. Gen. Ulysses S. Grant and Gen. Braxton Bragg, commanding generals of the Chattanooga Campaign.

Date	September 21 – November 25, 1863[1]
Location	Chattanooga, Tennessee

[hide]
Reopening the Tennessee River

Wheeler's Raid · Farmington · Brown's Ferry · Wauhatchie · **Chattanooga**

[hide]
Chattanooga –Ringgold Campaign

Wheeler's Raid · Farmington · Brown's Ferry · Wauhatchie · Lookout Mountain · Missionary Ridge · Ringgold Gap

The Chattanooga Campaign[2] was a series of maneuvers and battles in October and November 1863, during the American Civil War. Following the defeat of Maj. Gen. William S. Rosecrans' Union Army of the Cumberland at the Battle of Chickamauga in September, the Confederate Army of Tennessee under Gen. Braxton Bragg besieged Rosecrans and his men by occupying key high terrain around Chattanooga, Tennessee. Maj. Gen. Ulysses S. Grant was given command of Union forces in the West and significant reinforcements began to arrive with him in Chattanooga from Mississippi and the Eastern Theater.

After opening a supply line (the "Cracker Line") to feed his starving men and animals, Grant's army fought off a Confederate counterattack at the Battle of Wauhatchie on October 28–29, 1863. On November 23, the Army of the Cumberland under Maj. Gen. George H. Thomas advanced from the fortifications around Chattanooga to seize the minor high ground at Orchard Knob while elements of the Union Army of the Tennessee under Maj. Gen. William T. Sherman maneuvered to launch a surprise attack against Bragg's right flank on Missionary Ridge. On November 24, Eastern Theater troops under Maj. Gen. Joseph Hooker defeated the Confederates in the Battle of Lookout Mountain and began a movement toward Bragg's left flank at Rossville.

On November 25, Sherman's attack on Bragg's right flank made little progress. Hoping to distract Bragg's attention, Grant authorized Thomas's army to advance in the center of his line to the base of Missionary Ridge. A combination of misunderstood orders and the pressure of the tactical situation caused Thomas's men to surge to the top of Missionary Ridge, routing the Army of Tennessee, which retreated to Dalton, Georgia, fighting off the Union pursuit successfully at the Battle of Ringgold Gap. Bragg's defeat eliminated the last Confederate control of Tennessee and opened the door to an invasion of the Deep South, leading to Sherman's Atlanta Campaign of 1864

COMMON COUNCIL.

ADJOURNED SESSION—OFFICIAL PROCEEDINGS

IN COMMON COUNCIL, }
ADJOURNED SESSION, }
WEDNESDAY, March 8, 1865. }

On motion of Ald. Ennis, Ald. Bliss took the chair.
The roll having been called, the following members answered to their names as being present:
Alds. Whitney, McLaughlin, Wallace, Newnan, Ennis, O'Brien, Belknap, Wilson, Hathaway, Kelly, Kimball, O'Keeffe, Turton, Brady, Taylor, Sael, Bliss, Kalbfleisch, Fisher, Bulmer.

The reading of the minutes of the last meeting was dispensed with.

By Ald. Wallace—Deciding to repair a hole in Atlantic street between Henry and Hicks streets, adjacent to a manhole in the tunnel, at an expense not to exceed $25.

The resolution was adopted by the following vote:
Affirmative—Ald. Whitney, McLaughlin, Wallace, O'Brien, Belknap, Wilson, Hathaway, Kelly, Kimball, Turton, Brady, Taylor, Bliss, Kalbfleisch, Bulmer—15.
Negative—0.

Unanimous consent was granted.

Ald. Wallace moved that the Grading and Paving Committee be instructed to inquire into the condition of the tunnel in Atlantic street, and report what action should be had by the Board in relation thereto.

The motion was agreed to.

BELIEVES WILKES BOOTH IS ALIVE.

A Singular Story From One of the Assassin's Intimates.

CHICAGO, Ill., April 21.

The *Times* publishes a story from Birmingham, Ala., in which Louise Worcester, at one time a confidante of J. Wilkes Booth, the assassin of Abraham Lincoln, is credited with saying that Booth is not dead. She declares that in 1867, two years after Booth's supposed death, she received a letter without date or signature, but unmistakably in Booth's handwriting. This letter she says is still in existence. As to the probabilities of the man shot by Boston Corbett being Wilkes Booth, she points out that the body was closely guarded and secretly buried without an opportunity having been given for identification by any of those intimately acquainted with him. She believes that the man killed was one of the conspirators and that Booth made good his escape, but that in the excited and clamorous condition of the public mind it was thought best by the authorities, if they knew of the deception, to allow it to pass unchallenged, in order to allay the fever of excitement which the assassination had aroused.

Brooklyn's Notorious "Smoky Hollow" Gang

Police Battle River Pirates, ca. 1900

In the mid nineteenth century, the City of Brooklyn was booming from an economic point of view, but from a social perspective, it was an example of a Tale Of Two Cities (as was its sister metropolis, New York City). Vast numbers of poverty stricken recent immigrants were congregated into extremely overcrowded tenements immediately south of Atlantic Street (This tenement neighborhood was formerly a tidal wetland farm area). This area adjacent to the downtown Brooklyn waterfront, came to be known as "Smoky Hollow"; it was originally bound by Atlantic Avenue, Amity Street, Hicks Street and the waterfront. As time went on, the boundaries of Smoky Hollow expanded.

Smoky Hollow was a slum of crime and poverty that rivaled Manhattan's infamous Five Points. This notorious area existed for over half a century, starting in the 1860's. This center of assignation was presided over by the Smoky Hollow Gang, said to have been organized in 1867. They were a motley crew of cop killers, river pirates, smugglers, thieves, bootleggers and home invaders. They were led by the Mungerford brothers, Edward Glynn and Miles McPartland. For amusement, they are known to have nearly beaten to death a poor Organ Grinder, and slashed the throat of a musician who's tunes they didn't care for.

The smoky hollow gang's illegal activities were both tolerated and protected by "Boss" McLaughlin's City of Brooklyn political machine. McLaughlin's "machine" was said to have utilized the gang as its "enforcement" and "get out the vote" arm, via 6th Ward Alderman James Dunne. Dunne was described by a contemporary as a "prize fighter, ballot box stuffer and protector of thieves". This combination was no doubt patterned after the same sort of relationship that existed between Manhattan's Tammany Hall and gangs such as the Dead Rabbits.

Oddly enough, it appears the denizens of Smoky Hollow didn't make much use of firearms. Their weapon of choice was the straight razor, though they sometimes used a twelve pound cobblestone as a bludgeon.

Robert Diamond

SATELLITE SOLAR POWER STATION

A satellite equipped with extensive banks of solar cells is put into geo-synchronous orbit. This keeps it continuously above the same point of the earth where there is a receiving station. The sun's energy is absorbed by the banks of solar cells and is converted into electric energy. This energy is then converted to radio microwaves and beamed to the receiving station on earth. Here it is reconverted into electrical energy for transmission to homes, offices, farms and factories. Since the satellite is at altitudes much higher than the cloud levels, its reception of energy from the sun is uninterrupted by adverse weather conditions.

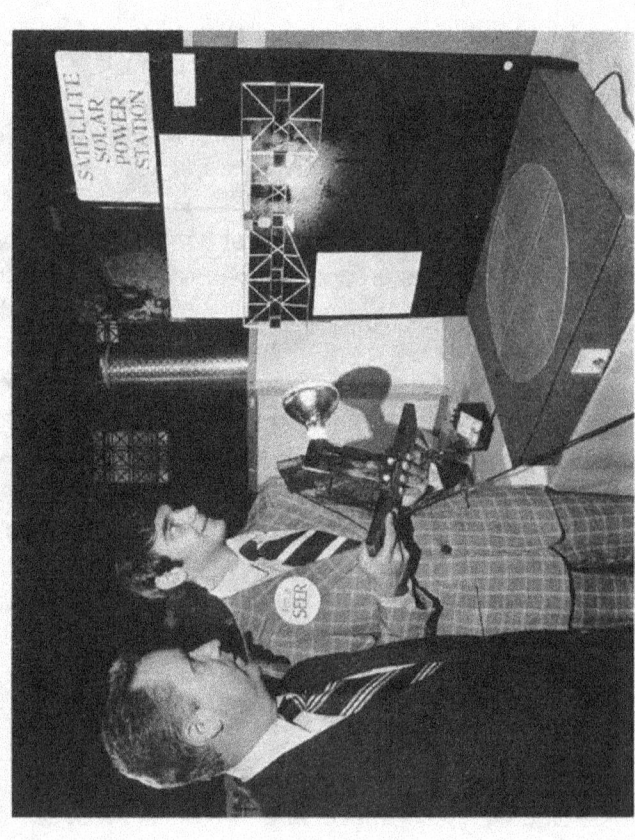

Robert Diamond shows his entry to NEF President Alan L. Smith.

Bob Diamond with National Energy Foundation President Allen Smith.

ALAN L. SMITH

ASSISTANT VICE PRESIDENT

Brooklyn Union Gas
195 MONTAGUE STREET, BROOKLYN, N.Y. 11201-3631
(718) 403-2525

Students in the News

Midwood is proud to congratulate those students who have received awards for their recent achievements in academics, service and athletics.

Marie Milford and Rudolf Rosefort received prizes of $50 in the Societe des Professeurs Francais Contest in the "native speaker" category...The Columbia Association of New York awarded a $50 savings bond to Anthony Voce for proficiency in Italian...The same Association awarded a medal to Janet Perrotta...Randy Roberts has received the District Attorney "Citation of Honor" Award in recognition of his "Progress Through Justice" in service to the school and community...Myron Diftler has won the Rensselaer Medal, awarded to a junior who is outstanding in Math and Science...Robert Diamond, an outstanding science student at Midwood, has won many awards, cash, and trips including first prize in SEER Science Contest, and a science award to be presented by Nelson Rockefeller. Midwood is extremely proud of Robert...Laurie Walker and Alison Kaluber won the Student Representative Contest sponsored by the Times College and School Service. Each received a cash award of $500.00.

Bob Diamond won a meeting with James R. Schlesinger, formerly Secretary of Defense and CIA Director under Presidents Nixon and Ford, then serving as America's first Secretary of Energy under Jimmy Carter. Schlesinger also served on the U.S. National Security Council, and as Chairman of the MITRE Corporation, until his death in 2014

President Gerald Ford meeting with the National Security Council in the Cabinet Room of the White House. Photographic negative showing President Ford seated at a table with William E. Colby, Director of CIA; Robert S. Ingersoll, Deputy Secretary of State; Henry Kissinger, Secretary of State; James R. Schlesinger, Secretary of Defense; William P. Clements, Jr., Deputy Secretary of Defense; George S. Brown, Chairman of Joint Chiefs.

"It has everything—mystery, adventure, history, and a delightful unsuspected ending. ... The unique tale of an American Sherlock Holmes." —*Seattle Times Magazine*

The COSGROVE REPORT

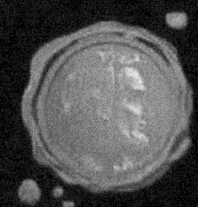

BEING THE PRIVATE INQUIRY OF A
PINKERTON DETECTIVE
INTO THE DEATH OF
PRESIDENT LINCOLN

AN ANNOTATED NOVEL PRESENTED BY
G.J.A. O'TOOLE

BROOKLYN HAS THE OLDEST SUBWAY IN THE WORLD

It Was Begun in 1836 Under Atlantic Avenue and Runs From South Ferry to Boerum Place, Being Part of the Route of the Brooklyn and Jamaica Railroad---Trains Actually Passed Through It, Preceded by a Man on Horseback---Later Used by Smugglers and Thieves.

ATLANTIC ST.   AVE.

COURT

BULLETIN No. 10

Locomotives of the Long Island Railroad.

By Inglis Stewart.

Railway and Locomotive Historical Society.

COPYRIGHT 1925

---

3. Hicksville.

This was built by the proprietors of the Locks & Canals Co., at Lowell, Massachusetts, Geo. W. Whistler, Manager, and in 1836 began service on the L. I. R. R. Whether it came direct to the L. I. R. R., or had been in service elsewhere earlier than 1836 I have been unable to ascertain, but am inclined to think it came direct from the Lowell Shop. It was scheduled as embraced in the State of Michigan mortgage. The only details that have been discovered are the following: Weight 7 tons, cylinders 11x16; 1 pair drivers 5 ft. in diameter. It seems not to have run after 1848, but was owned as late as 1853. It was then deemed not worth repairing. Its disposition could not be ascertained.

Long Island Rail Road locomotives from 1836 through 1861:

A roster from the beginning of the LIRR through the end of operation in the Atlantic Avenue Tunnel.

NAME	WHEEL ARRANGEMENT	BUILDER C/N	DATE	CYLINDERS	DRIVERS	WEIGHT	NOTES
Ariel	4-2-0	Baldwin 19	11/28/1835	10x16	54	7 tons	
Post Boy	4-2-0	Baldwin 26	3/23/1836	10x16	54	7 tons	
Hicksville	2-2-0	Locks & Canals	1836	11x16	60	7 tons	
John A. King ex Taglione	2-2-0	Poughkeepsie	1838	11x16	60	7 tons	
Chichester	4-4-0	Baldwin 143	1840	12x16	48	11 tons	Rebuilt by Campbell from single driver
Crabb	4-2-0	Baldwin 141	1840	13½x16	42	11 tons	
Brooklyn (Rebuilt From Crabb)	0-6-0	Rebuilt LIRR	1852-1853	13½x16	39	12 tons	
Brooks	4-4-0	Rogers 53	4/25/1844	11½x22	60	15 tons	
Fisk	4-4-0 (4-2-2?)	Rogers 58	8/10/1844	12x20	64	16 tons	
James H. Weeks ex Edwin Post	2-4-0	Baldwin	ca. 1843	13½x16	60	14 tons	
Elihu Townsend	4-4-0	Norris	1844	12½x20	60	17 tons	
Derby	4-4-0	Hinkley 24	9/13/1844	13½x20	66	16 tons	
Boston	4-4-0	Hinkley 46	7/29/1845	14½x20	54	17 tons	
Little Renamed Fanny	4-2-0 (4-2-2?)	Norris	1845	12½x20	69	15 tons	
Ruggles	4-2-0 (4-2-2?)	Norris	1845	12½x20	69	15 tons	
New York	4-4-0	Rogers 71	8/22/1845	12½x20	72	17 tons	
Moses Maynard	4-4-0	Rogers 270	7/24/1851	14x20	66	20 tons	
Long Island	4-4-0	Rogers 311	4/3/1852	14x22	72		
Peconic	4-2-0	Norris	1853	12½x24	60	16 tons	
Wyandank	0-6-0	Baldwin 555	10/1853	16x24	40	21 tons	
Rebuilt	4-4-0			16x20	48		
Montauk	4-4-0?	Swinburne	1854	15x20	60	20 tons	
Orient	4-4-0	Rogers 501	6/24/1854	14x20	66	22 tons	
Atlantic	4-4-0	Rogers 578	5/11/1855	14x20	66	22 tons	
Nebraska*	4-4-0		1856				*Test, Coal Burner, Boardman Boiler
Pacific	4-4-0	Rogers	5/23/1857	14x20	66	22 tons	
Phoenix	4-4-0	Rogers	1/30/1860	15x22	60	25 tons	
Nassau	4-4-0	Norris	12/1860	14x24	60	24 tons	

643-2051

HOWARD GOLDEN
PRESIDENT

The City of New York
President of the Borough of Brooklyn

BOROUGH HALL
BROOKLYN CIVIC CENTER
BROOKLYN, N.Y. 11201

June 12, 1980

Mr. Robert Diamond
599 East 7th Street
Brooklyn, New York 11218

Dear Mr. Diamond:

    Ms. Mary Taintor, of my staff, has advised me that you have done extensive research on a train tunnel under Atlantic Avenue which contains a train from 1830 with a wood burning engine.

    Given your deep involvement in Brooklyn history, I would like to invite you to join my History Advisory Committee. The History Advisory Committee is comprised of Brooklynites active in local history and local historical societies. By sponsoring projects to promote Brooklyn history and by creating a link among our many local history societies, the Committee focuses community attention on our fascinating heritage.

    I have requested Mr. Donald Simon, Chairperson of the History Advisory Committee to write to you inviting you to the Committee's next meeting.

    We look forward to working with you on promoting Brooklyn history.

Sincerely,

Howard Golden

MEMO From HERBERT M. KASS, P.E.
DISTRIBUTION ENGINEER
Department of Water Resources
Bureau of Water Supply
Brooklyn, N.Y. 11201

7/28/80

This will certify that Mr. Robert Diamond and a Water Supply crew investigated an old abandoned tunnel under Atlantic Avenue at Court Street.

A manhole cover was removed at that location and the tunnel was photographed by Mr. Diamond.

A Tunnel That Can Keep a Secret

Robert Diamond exploring railroad tunnel that was sealed off beneath Brooklyn's Atlantic Avenue around 1860. An accumulation of dirt near the tunnel's only entryway, a manhole, left about three feet of space under the brick ceiling.

Bootleggers distilled their spirits in the dark cavern below Atlantic Avenue, according to the Brooklyn rumor-mongerers of 1911. Murderers buried the remains of victims there, it was said in 1936. And during the two world wars, some Brooklynites swore that a German spy headquarters lay beneath their feet.

So it was with a mixture of historical curiosity and Hardy Boy spunk yesterday that Robert Diamond, a 20-year-old Brooklyn engineering student and railroad buff, descended into the 136-year-old railroad tunnel beneath Atlantic Avenue that had been sealed and abandoned at the beginning of the Civil War.

After researching the tunnel's past for seven months, Mr. Diamond — whose apartment in the Kensington section is adorned with chunks of railroad tracks and drawings of old locomotives — approached the office of Borough President Howard Golden recently about conducting a journey under Atlantic Avenue.

It would be the first attempt to explore the tunnel since it had been inspected in 1941 for possible use as an air-raid shelter. Officials grudgingly agreed, and other city agencies said they would provide safety equipment.

The Descent Begins

"I've heard rumors that Murder Inc., the Mafia, Judge Crater and everyone but Amelia Earhart has been in there," Mr. Diamond said before he descended. "I want to go in and see it for myself."

He got only 12 feet. City safety experts discovered carbon monoxide in the tunnel, and called off the expedition. They warned those gathered for the occasion that any venture inside could be extremely dangerous and should not be attempted by amateurs.

"The tunnel has generated so much mystery and fantasy because people question whether it is there at all," said Donald E. Simon, chairman of the Brooklyn Borough President's History Advisory Committee, who was among those looking on as Mr. Diamond readied himself.

"Despite what people say, the tunnel has never been lost, and people have been inside over the years," Mr. Simon said. "But no one has gone in to investigate it from a historical perspective. In its time, it was one of the longest railroad tunnels in existence, a significant engineering feat."

Clad in work boots and short pants, Mr. Diamond jumped down into the tunnel entryway, a five-foot-deep manhole at Atlantic Avenue and Court Street, as a dozen city employees and reporters awaited the arrival of equipment to test for poisonous gases.

Digging for Artifacts

While a machine provided by the city's Department of Environmental Protection poured in fresh air, Mr. Diamond sat a few yards from the manhole entrance, on a dirt floor about three feet under the arched brick ceiling of the tunnel, digging feverishly for artifacts. Off in the distance, beyond the range of his flashlight, he surmised, the dirt sloped down to the floor, allowing room for explorers to wander freely.

Mr. Diamond remained inside for an hour and a half, until city workers tested the air and found it unsafe, and he found himself, dirty and disappointed, on the street above the closed manhole.

The half-mile tunnel was completed by the Long Island Rail Road in the last months of 1844. According to railroad records, it is 21 feet wide and 18 feet high, with stone walls 6 feet thick. Trains traveled through the tunnel from Columbia Street to Court Street until about 1860, bringing produce and passengers from Long Island on a line from Jamaica to the Brooklyn waterfront.

Closed by the Competition

The tunnel was sealed by order of the State Legislature, Mr. Simon said, because Brooklyn merchants feared the rail line would bring increased prosperity to other areas of Long Island at Brooklyn's expense.

In the following years, the dark netherworld became a focus of borough folklore, if not a place of treasure and death. In 1911, The Brooklyn Daily Eagle reported tales of enormous rats and a distilling operation with a pipe connected to an Atlantic Avenue barroom.

"It is known the tunnel has seen gunplay," the paper reported, "and that battles have been fought in smugglers' lairs.

"The popular impression prevails today that it would be a good deal safer to go down into its darkness armed," the story continued, adding that the tunnel might be "still seriously guarded by desperate men."

LOCKHEED CORPORATION
BURBANK, CALIFORNIA 91520

NEW YORK OFFICE
(212) 371-3838

SUITE 2801
375 PARK AVENUE
NEW YORK, N.Y. 10152

September 4, 1980

Mr. Robert Diamond
599 East 7th Street
Brooklyn, New York 11218

Dear Mr. Diamond:

    Thank you for your letter and enclosed material. I have forwarded them on to the appropriate personnel department and you should be hearing from them shortly. Your background is impressive and I congratulate you on your various achievements.

    I hope you are successful at making another attempt on the tunnel and look forward to reading about it in the papers again.

Sincerely,

Marion Archard

Marion Archard
Manager

Junk those myths about rail tunnel

By DAVID HARDY

The mystery and myth surrounding the recently rediscovered Atlantic Ave. railroad tunnel has turned out to be just so much junk, according to a Brooklyn College student and two scientists who have explored the legendary underground passageway built in 1844.

"We found lots of railroad spikes, some empty whisky bottles and the skeleton of a dog that apparently got trapped there. But I was disappointed because there was no steam locomotive," said Robert Diamond, 21, a student at City College who organized and led Tuesday's exploration of the 1,700-foot passageway.

Once used by the old Brooklyn and Jamaica Railroad to bring farm produce from Long Island to Brooklyn's waterfront, the tunnel ran from Boerum Place to Willow Place before it was sealed up and largely forgotten in 1861.

According to local legend, the tunnel was rumored at various times to harbor eight-foot man-eating rats, poisonous snakes and illegal whisky stills. It was rumored to be a dumping ground for victims of Murder Inc., and a hiding place for German spies during the world wars.

Diamond said he first learned of the mysterious passageway from a local radio program. Curious, he then set about digging into city records until he found blueprints showing that the tunnel's only entrance was a manhole at Atlantic Ave. and Court St.

As things happened, Diamond's first planned exploration of the tunnel was called off last August when city engineers warned that the chamber might contain noxious gases.

During the months of delay until various city agencies got around to giving their okay for the expedition, speculation about the mysteries of the passageway heightened. That is, until Tuesday when Diamond, along with two archeologists and a Brooklyn Union Gas Co. engineering team, finally got the okay to descend into the tunnel—only to discover the tunnel contained lots of very ordinary junk.

The two archeologists—Eugene Sterud, executive director of the Archeological Institute of America, and Joel Klein of the Manhattan-based Envirosphere Inc.—said they spent several hours exploring the tunnel and found nothing of "scientific importance."

"Probably, the thing to do is simply record that it's there and seal it back up so that it isn't destroyed and doesn't present a hazard to anyone," Sterud said.

There is so a locomotive in there, he says

By ALBERT DAVILA

There is definitely an old steam locomotive located in a portion of the mysterious Atlantic Ave. tunnel in Downtown Brooklyn which is now sealed, asserted yesterday a man who last saw the wood-burning engine more than 40 years ago.

"We used to play with it when I was a kid," said Juan Vega, a 55-year-old Brooklyn seaman who in his youth used to live in a tenement building at 64 Atlantic Ave.

"There used to be a hole in the basement of the building and we made it bigger and went into the tunnel," Vega continued. "We used to take penny candles down there and play."

The engine, he said, was atop some wooden ties in a portion of the tunnel that extends from Hicks St. to Columbia St.

"It looked older than the engines we saw in the cowboy movies," Vega said. "It used to have a brass plaque on its instrument panel which we used to polish every now and then. Engraved on that plaque was the name of the manufacturer and the date—all of which I've long forgotten."

Daily News, Friday, September 11, 1981

N.Y. Daily News — 1981

View of west side of concrete wall before installation of wood stair. Note chain ladder and access opening

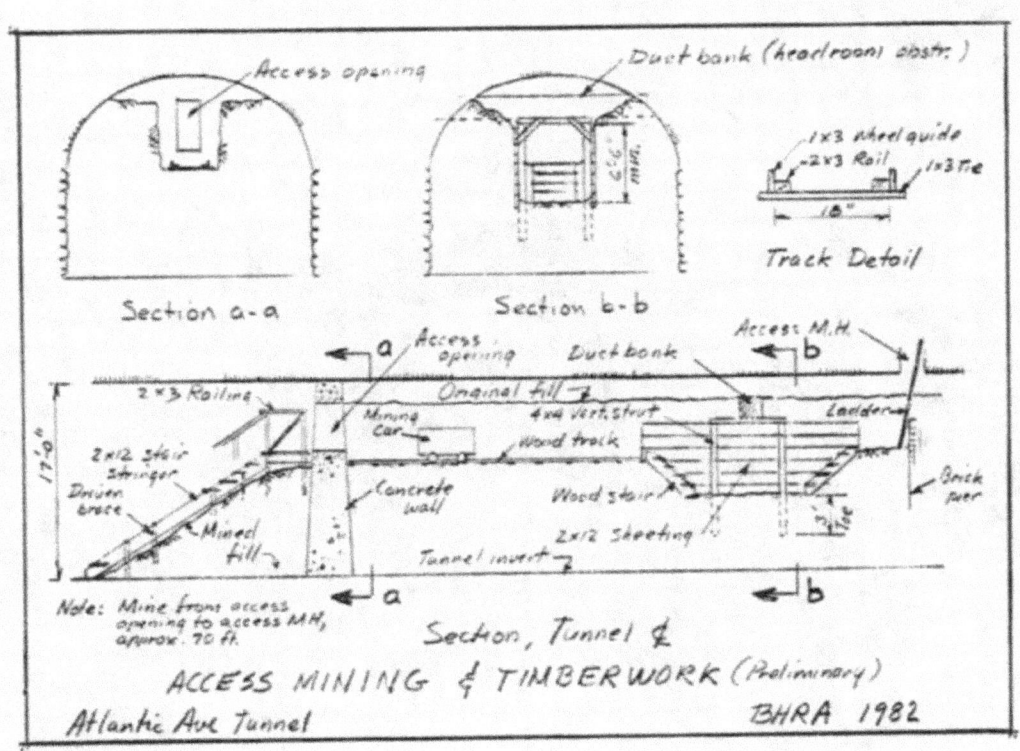

Temporary track and mining car used in trenching phase.

Right- Bob outside the manhole on Atlantic avenue shortly after his discovery.

Left: BRHA volunteers improving access in 1982.

History lurking underground

Tour the world's oldest subway tunnel under busy boro street

BY DENISE ROMANO

HE'S GOT tunnel vision.
Bob Diamond gives tours of the 165-year-old Atlantic Ave. subway tunnel once a month.

The discovery of the world's oldest subway tunnel began in 1979, when Diamond was an electrical engineering student.

"I read about missing pages from a book written by John Wilkes Booth," he said. "Those pages were reportedly hidden in a subway tunnel, in New York."

After lots of research in his spare time, Diamond found the Atlantic Ave. tunnel. No missing pages, though.

"I used a map from 1911 to find this tunnel," he said, adding that after much negotiation with the city, he received permission to take the big trip underground.

Diamond was so fascinated, he started giving tours of the half-mile long tunnel in 1982.

The tunnel was built by hand labor, in just seven months in 1844, for Long Island Rail Road trains, which did not have brakes good enough to operate on city streets.

By running the trains underground, delays on the street were eliminated.

The tunnel tour is not for anyone who is out of shape.

It starts with a climb down through a manhole at Atlantic Ave. and Court St.

"During the tour, I give its history and tell folklore of it," Diamond said.

"Bootleggers and smugglers were down there, as well as river pirates. Murder Inc. supposedly dumped bodies there in the 1930s. It's so full of history."

For more info, visit www.brooklynrail.net or call Diamond at (718) 941-3160.

Photos by Nicholas Fevelo

A tour group in the 1844 Long Island Rail Road tunnel under Atlantic Ave. listens to guide Bob Diamond talk about the history of the tunnel, which was built with rock (L) from Manhattan.

Highly popular Tunnel Tours and Special Events, conducted regularly from 1982-2010

Bob Diamond, August, 2007 tunnel Tour. Photo: Justin N. Lane

Your Borough

TUNNEL GETS LOVE

Guinness world record for an 1844 LIRR relic

BY MIKE McLAUGHLIN
DAILY NEWS WRITER

IT'S DARK, dirty and neglected, but a train tunnel beneath Atlantic Ave. is basking in glory after being dubbed the world's oldest in the new "Guinness World Records."

The half-mile long passage was built in 1844 — making it 60 years older than the city's subway system, which went on sale Sept. 15.

Though Guinness calls it a subway tunnel, the passageway actually was part of a network for trains that ran to Greenport, L.I.

The tunnel, which reaches from Columbia St. almost to Boerum Place in Cobble Hill, made the area safer for pedestrians.

"The tunnel was the first in the world built underground in order to improve urban congestion, public safety and rail operations," said the entry from the 2011 Guinness book.

In other parts of Brooklyn farther east, tracks rose above ground and ran on the street, creating sometimes deadly hazards.

The underground route was sealed off in 1861 and was practically forgotten until transit buff Bob Diamond of Kensington rediscovered it in 1980, when he was 19.

"Now, the tunnel finally has the recognition that it deserves," said Diamond, who founded the Brooklyn Historic Railway Association, which gives monthly tours of the subterranean passage.

A Guinness researcher contacted Diamond earlier this year for documents about the tunnel's history, he said.

But Diamond only found out that it landed in the record book this past Sunday from a volunteer with his tour group.

"I jumped about 3 feet in the air," Diamond said. "I've been advocating for this tunnel for the last 30 years."

Many world records are submitted by readers to Guinness officials for verification, but a spokeswoman said the book's staff created the subway tunnel category themselves.

Today, the tunnel is city-owned property, but it was immaculately built by the Long Island Rail Road, said railroad historian David Morrison.

The arched ceiling is 17 feet tall and made of brick. The walls, 21 feet apart, feature large stones.

But the tunnel's craftsmanship was lost on train riders who couldn't see it because it was too dark.

"The people would see nothing but darkness," said Morrison. "For something that nobody was going to see, the work was extraordinary."

Today's MTA straphangers have nothing to complain about compared to those early riders, Morrison said, adding: "In those days, you got dirty from riding trains with cinders from the locomotives."

Bob Diamond, at the entrance leading to 1844 Long Island Rail Road tunnel beneath Atlantic Ave. in Cobble Hill, is proud curiosity he rediscovered is getting just due. Photo by Andrew Theodorakis/Daily News

AT A GLANCE

DOWNTOWN — Artist Ilana Halperin, anthropologist Karen Holmberg and art historian Andrew Patrizio to offer an evening of lectures that examine the status of volcanoes in contemporary culture tomorrow at 7 p.m. at Triple Canopy, 177 Livingston St. The Hand Held Lava

The team will present field notes from volcanic sites across the world and will attempt to make sense of our ongoing desire to explore volcanoes. There is a $5 donation. For info, visit canopycanopycanopy.com.

MARINE PARK — Explore the differences between song sparrows and swamp sparrows, and take a look at the other little brown birds in and around Marine Park at noon on Sunday.

Meet at the Salt Marsh Nature Center, 332 Avenue U. It's free and fun for all ages. For more info, visit www.nyc.gov/parks/rangers.

FORT GREENE — MoCADA presents a film screening and discussion on "Road to Resistance" today from 6:30 p.m. to 8 p.m. at 80 Hanson Place.

The film explores the global citizens" movement that took on South Africa's apartheid regime. Call (718) 230-0492.

BAY RIDGE — The Sean Casey Animal Rescue Benefit will be held at 7 p.m. Sunday at the Wicked Monk Irish Pub, 8415 Fifth Ave.

Live performances include Two for the Road at 7 p.m., Ghosts of Eden at

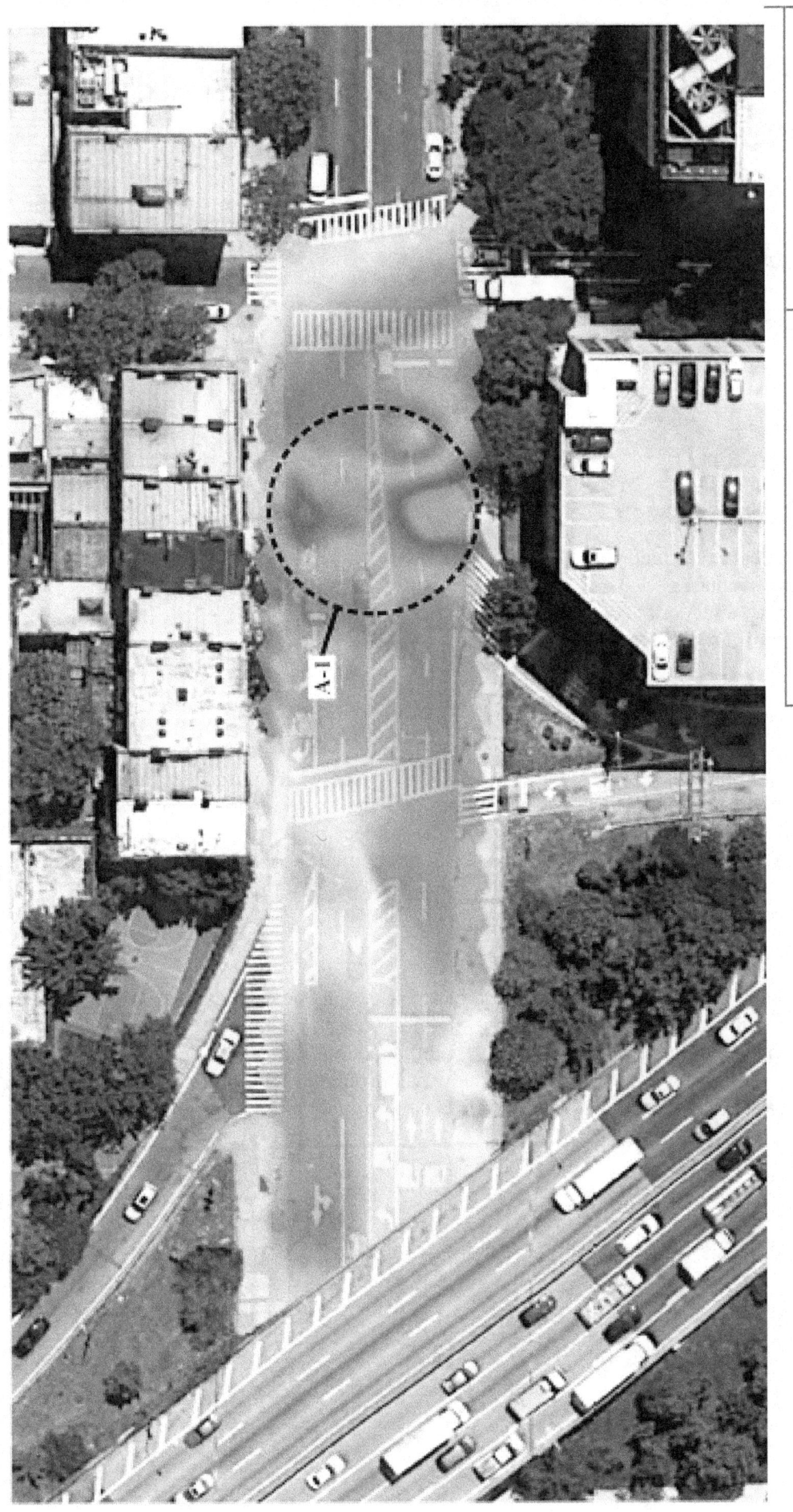

APPENDIX C

BRINKERHOFF
ENVIRONMENTAL SERVICES, INC.

1913 Atlantic Avenue, Suite R5
Manasquan, New Jersey 08736
Tel: (732) 223-2225
Fax: (732) 223-3666

January 18, 2011

Janine Hildebrand, EIT
S. Harris, Ltd.
2601 Pennsylvania Avenue, Suite Eight
Philadelphia, PA 19130

Re: Geophysical Investigation Report
     Atlantic Avenue
     Brooklyn, New York
     Brinkerhoff Project No. 10BR194

Dear Ms. Hildebrand:

Brinkerhoff Environmental Services, Inc. (Brinkerhoff) is pleased to present the following summary report of the Geophysical Investigation conducted on January 11 and 14, 2011 at the above-referenced property (herein referred to as the subject property). Refer to Figure 1 – Site Location Map. Electromagnetic induction, electromagnetic soil conductivity, total field magnetics and ground penetrating radar (GPR) were employed for the investigation.

Introduction

On January 11 and 14, 2011, Brinkerhoff conducted a geophysical investigation on the subject property. The purpose of the geophysical investigation was to evaluate the potential presence of subsurface anomalies indicative of a buried 19th century locomotive and associated artifacts. The subject property is currently an active urban roadway surrounded by buildings to the north, south and east, while a large steel overpass (I-278) borders the subject property to the west.

APPENDIX C

Janine Hildebrand, EIT
Re: Geophysical Investigation Report
      Atlantic Avenue
      Brooklyn, New York
      Brinkerhoff Project No. 10BR194
January 18, 2011

GEOPHYSICAL CONCLUSIONS

On January 11 and 14, 2011, Brinkerhoff performed a geophysical investigation in open and accessible areas of the subject property. Limitations encountered during the investigation included the presence of possible subsurface utilities, metallic light poles, suspect concrete road bedding, vehicles, vegetation, snow, refuse piles, adjacent structures and the I-278 overpass. Please note that Electromagnetic Induction, Terrain Conductivity, Total Field Magnetics and GPR measurement are remote sensing methods and in some instances, due to interference or other geophysical limitations, do not reveal data which may be indicative of subsurface anomalies. One (1) large subsurface metallic anomaly was identified extending across Atlantic Avenue and encompassing both the west bound and east bound roadway. Brinkerhoff was able to estimate the metallic anomaly's length at 20 feet based upon the response of the G-859 data. The anomaly was outlined in white spray paint in the field and is noted on the attached Figure 3 and Figure 4.

This report has been prepared and is respectfully submitted by

BRINKERHOFF ENVIRONMENTAL SERVICES, INC.

January 18, 2011

MATTHEW D. POWERS   Date
Director of Geophysical Services

RE: Atlantic Avenue Tunnel - Proposed Boring Location Plan

From: **Matt Powers** (mpowers@brinkenv.com)
Sent: Fri 1/28/11 5:12 PM
To: 'lynn rakos' (lrakos@hotmail.com); janine@sharrisltd.com
Cc: sam@sharrisltd.com; 'Laura Brinkerhoff' (lbrinkerhoff@brinkenv.com)

Lynn,

Based upon Brinkerhoff's interpolation of the geophysical data, the "heart of the anomaly radiates around the center of the street and slightly skewed towards the south. Brinkerhoff is finding it extremely difficult to determine if the large geomagnetic response is due to one (1) very large subsurface anomaly or a couple large anomalies. The actual anomaly is not as large as the image map portrays it. The is due to the anomaly containing a great deal of magnetic energy resulting in an elongated visual response. It is conceivable that the suspect locomotive is located between the middle and south sides of Atlantic Ave. and a separate smaller anomaly is located on the northern side of Atlantic Ave. Based upon Brinkerhoff's interpretation of the geophysical data, there is no question that something(s) metallic is buried under Atlantic Ave., its just a matter of what and in what orientation.

I as well found it odd for the signature to extend past the tunnels walls however; I am going to have to revert back to the shear size of the magnetic response as I mentioned above.

The blip to the east represents a magnetic response and may represent additional artifacts however; I do not feel that this response is large enough to represent a locomotive.

Brinkerhoff does not believe that the response is due to any surficial utilities that InfraMap identified.

Lynn, I hope I answered your questions but if I haven't, please feel free to contact me at anytime.

I hope you have a happy weekend as well.

Thank you,

Matthew D. Powers
Director of Geophysical Services

"Blending fantasy and factual accuracy, *Hubert Robert*'s (22 May 1733 – 15 April 1808) views of classical and contemporary architecture were immensely popular during his lifetime. Robert was best known for his paintings of ruins. His immense, crumbling monuments of the past earned him the nickname, "Robert des Ruines" (Robert of the Ruins)." **According to W.H. Adams "Hubert Robert's paintings both recorded and inspired".**

Robert: Felsengrotte mit antiker Architektur

Sources: http://urbanism101.rssing.com/chan-7185228/all_p3.html
https://en.wikipedia.org/wiki/Hubert_Robert
http://it.wahooart.com/Art.nsf/Diaporama?Open&A=Hubert Robert

"The 9th Ave. station, lower level, served the original BMT Culver and West End Lines from 1919 to 1954. The IND connection to the Culver Line was made at Ditmas Avenue in 1954, rerouting Culver trains via the new IND subway. The section of track along 39th Street between Ninth Avenue and Ditmas Avenue operated as a shuttle, using the lower level platforms at Ninth Ave. These lower platforms have been disused since the shuttle operation was terminated on May 13, 1975".

Source: http://www.nycsubway.org/wiki/Station:_9th_Avenue_(West_End)#Lower_Level

Rutherford Atom – Circa 1911

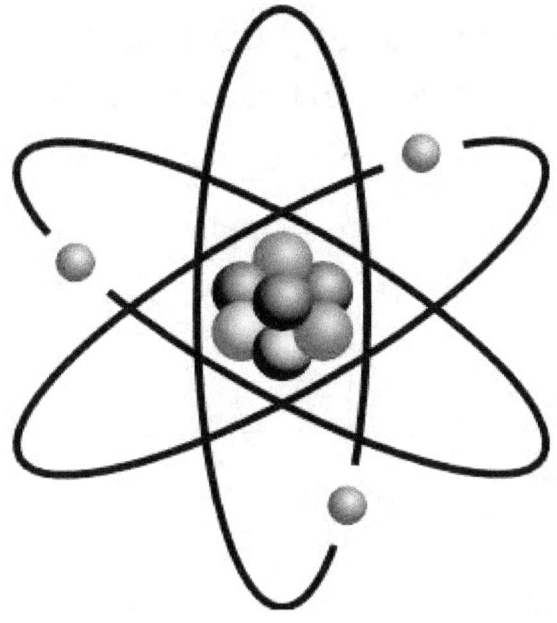

Schrodinger- Heisenberg Circa 1926 - 1927

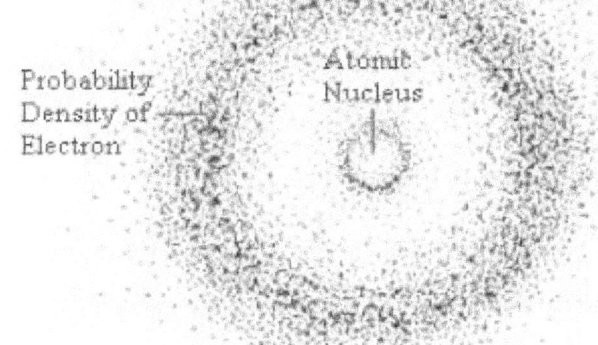

http://www.techrepublic.com/article/how-a-physics-discovery-85-years-in-the-making-could-change-how-we-build-electronics/

How a physics discovery 85 years in the making could change how we build electronics

Researchers at Princeton University have observed the **Weyl fermion, a massless particle first theorized 85 years ago**. Here's how the discovery could change the way we make electronics.

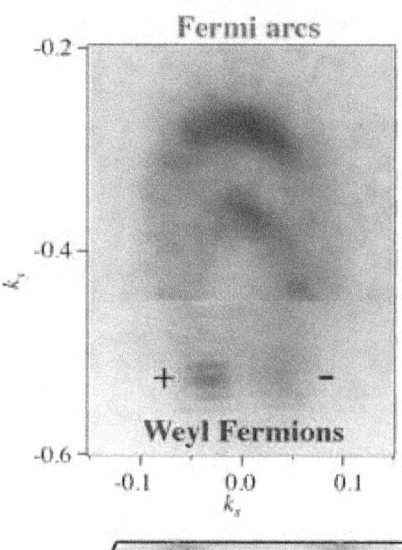

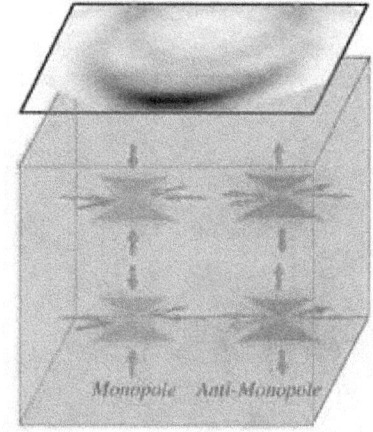

So What Happens when "m = 0" ?

Newton: $F = ma$

Einstein: $E = mc^2$

 You can throw these out

-BUT-

Lorentz $F = q[E + (v \times B)]$

Ohm: $E = IR$

 These still work

Concepts hitherto fore considered "impossible science fiction" may now possibly be reduced to a question of finding creative engineering solutions

Thu, Sep 10, 2015

http://www.businessinsider.com/everything-you-need-to-know-about-the-large-hadron-collider-2015-9

Scientists built the most powerful physics machine on earth to study the fate of our universe — and it may break the laws of physics

-Dark Matter
-Dark Energy
-Worm Holes

Magnetic Wormhole Created in Lab

Device acts like a wormhole, as if the magnetic field was transferred through an "extra special dimension"

By Tia Ghose and LiveScience | August 21, 2015

Ripped from the pages of a sci-fi novel, physicists have crafted a wormhole that tunnels a magnetic field through space.

"This device can transmit the magnetic field from one point in space to another point, through a path that is magnetically invisible," said study co-author Jordi Prat-Camps, a doctoral candidate in physics at the Autonomous University of Barcelona in Spain. "From a magnetic point of view, this device acts like a wormhole, as if the magnetic field was transferred through an extra special dimension."

©iStock.com

The idea of a wormhole comes from Albert Einstein's theories. In 1935, Einstein and colleague Nathan Rosen realized that the general theory of relativity allowed for the existence of bridges that could link two different points in space-time. Theoretically these Einstein-Rosen bridges, or wormholes, could allow something to tunnel instantly between great distances (though the tunnels in this theory are extremely tiny, so ordinarily wouldn't fit a space traveler). So far, no one has found evidence that space-time wormholes actually exist. [Science Fact or Fiction? The Plausibility of 10 Sci-Fi

PHYSICS

"Spooky Action at a Distance" Confirmed by New Quantum Experiment

March 31, 2015 | by Janet Fang http://www.iflscience.com/physics/einsteins-spooky-action-distance-confirmed-new-quantum-experiment

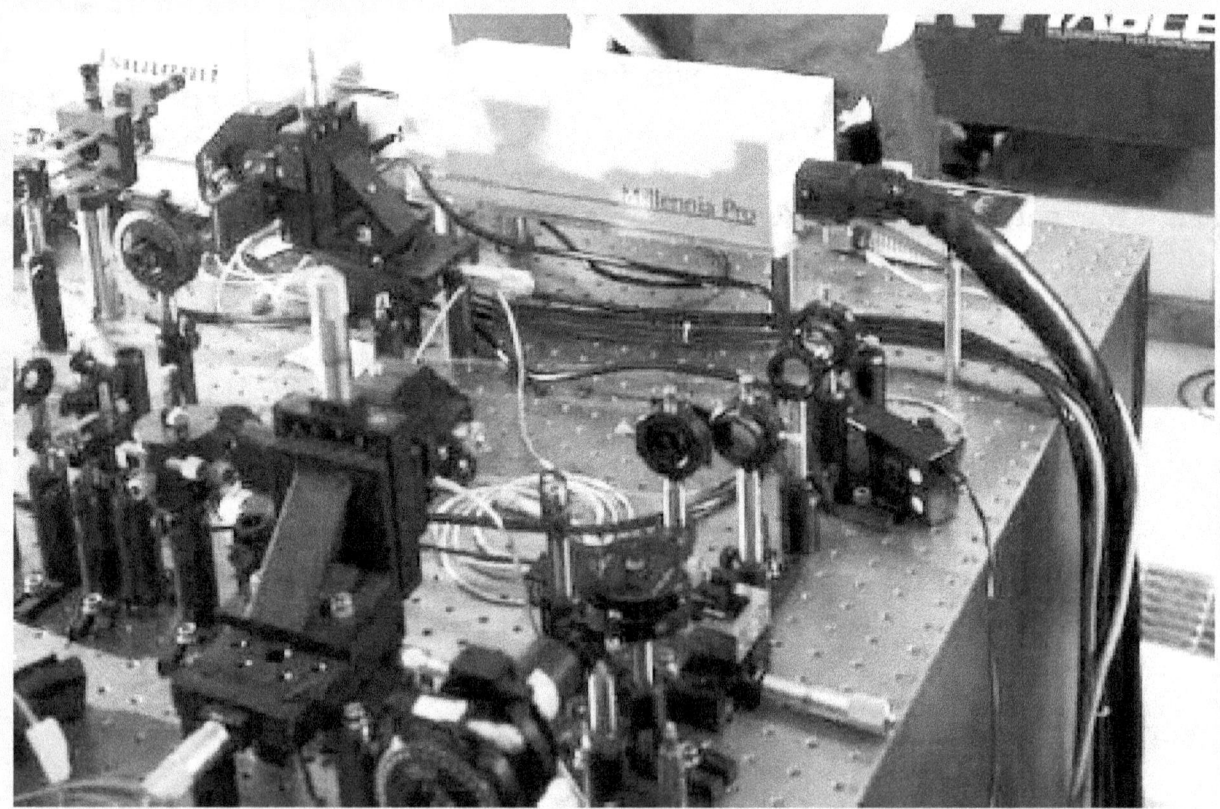

Photo credit: Griffith University

Albert Einstein may have been the greatest mind of the 20th century, but the great physicist famously disliked some of the weirder implications of quantum physics. Now, nearly a century after his protests, physicists may have proven one of the points that he doubted the most.

According to quantum mechanics, a particle can be described as a wave that spreads out over a great distance. Yet the particle is still just one particle. You can't detect it in two places at once. When physicists observe the particle in a particular location, they say that the wave function—the mathematics that describes how a particle could be in multiple places at once—has collapsed.

Einstein could not accept this. Or, at least, that he thought the quantum mechanics of his day could not adequately explain it, referring to the phenomenon with the now-iconic phrase "spooky action at a distance." But in new research published in *Nature Communications*, Griffith University's Howard Wiseman and colleagues use a single particle to show that the wave function really does collapse in this strange way. In so doing, their work backs up years of research into quantum entanglement, in which particles are connected in a mysterious way even when separated, so that observing or affecting one instantly affects the other.

Previous experiments had tested quantum entanglement with two particles, but the researchers wanted to get at Einstein's claim by entangling a single photon of light. They did this by firing a beam of photons into a splitter that cut each photon in two, sending half of the light to one lab and half to another lab.

Using a finely tuned homodyne detector—a tool used to measure the waves of these particles—Lab A tried to look for its photon and measure its phase. So did the scientists in Lab B. They found that if the Lab A researchers had detected the photon, then the Lab B researchers did not, and vice versa. Plus the photon state that Lab B detected depended upon what Lab A detected. That's exactly what you'd expect if the single split photon were entangled.

"Einstein's view was that the detection of the particle only ever at one point could be much better explained by the hypothesis that the particle is only ever at one point, without invoking the instantaneous collapse of the wave function to nothing at all other points," Wiseman says in a news release. "Through these different measurements, you see the wave function collapse in different ways, thus proving its existence and showing that Einstein was wrong."

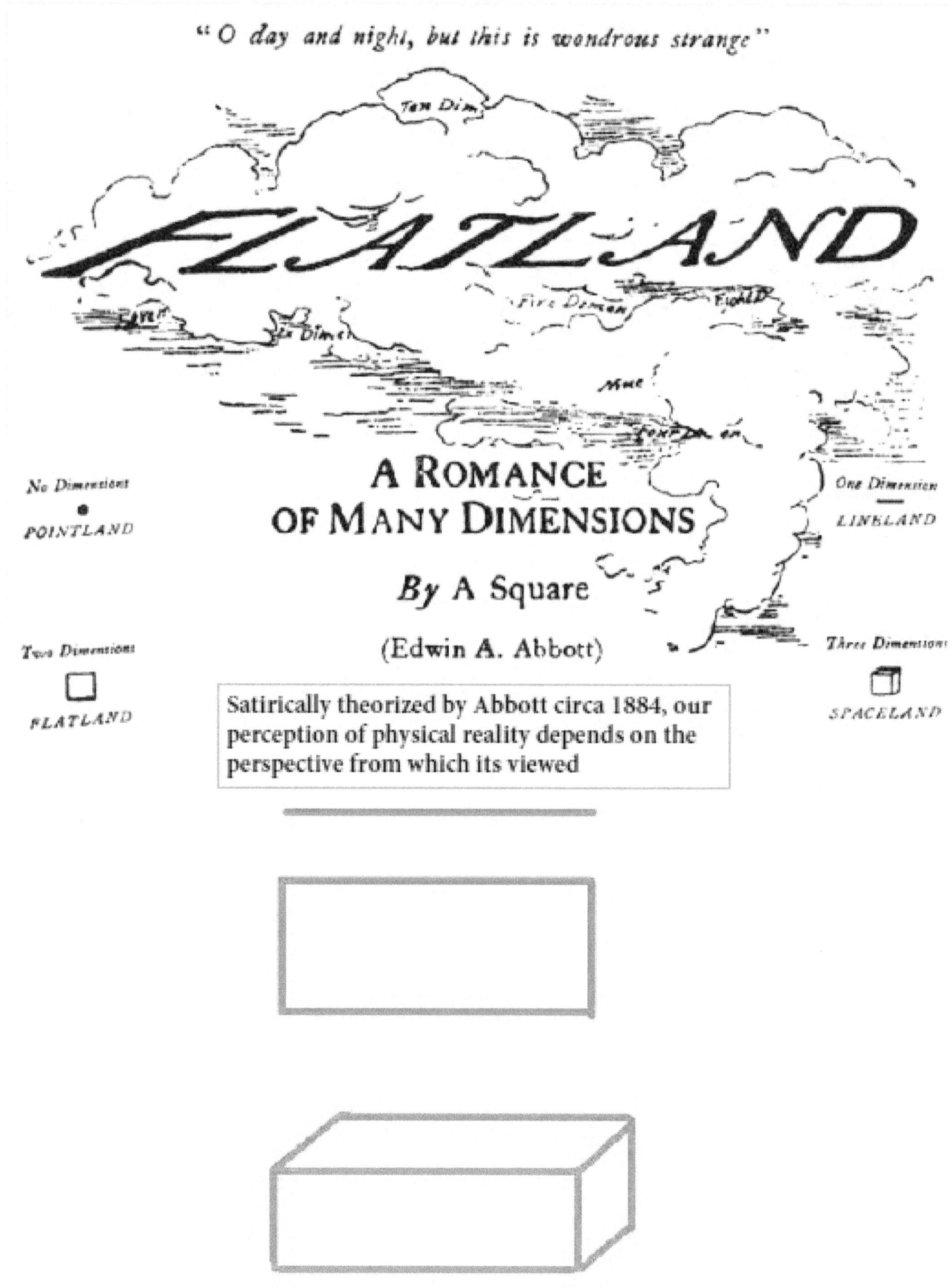

The Technology

Creating high-resolution full-color moving holograms in 3-D

4 February 2015

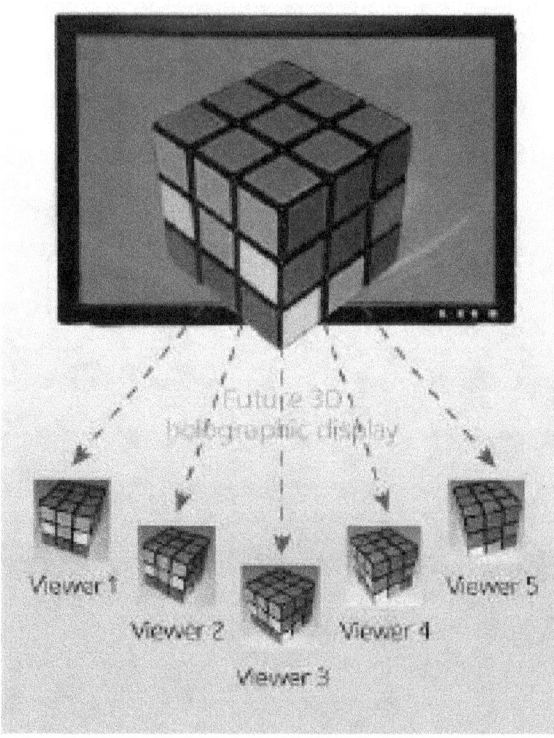

A new way of streaming high-resolution, full-color full-parallax three-dimensional (3D) hologram videos may have applications in the entertainment and medical imaging industries. Credit: A*STAR Data Storage Institute

Holograms are considered to be truly 3D, because they allow the viewer to see different perspectives of a reconstructed 3D object from different angles and locations (see image). Like a photograph, a hologram contains information about the size, shape and color of an object. Where holograms differ from photographs is that they are created using lasers, which can produce the complex light interference patterns, including spatial data, required to re-create a complete 3D object.

However, generating high-resolution, moving holograms to replace current 3D imaging technology has proved difficult. To enhance the resolution of their holographic videos, Xuewu Xu and colleagues at the Data Storage Institute in Singapore used an array of spatial light modulators (SLMs).

"SLMs are devices used in current two-dimensional projectors to alter light waves and generate projections," explains Xu. "In a 3D holographic display, SLMs are used to display hologram pixels and create 3D objects by light diffraction. Each SLM in our system can display up to 1.89 billion hologram pixels every second, but this resolution is not high enough for a seamless large video display."

Three-dimensional (3D) movies, which require viewers to wear stereoscopic glasses, have become very popular in recent years. However, the 3D effect produced by the glasses cannot provide perfect depth cues. Furthermore, it is not possible to move one's head and observe that objects appear different from different angles—a real-life effect known as motion parallax. Now, A*STAR researchers have developed a new way of generating high-resolution, full-color, 3D videos that uses holographic technology.

To address this challenge, Xu and his team divided every frame of their hologram video into 288 sub-holograms. They then streamed the sub-holograms through 24 high-speed SLMs stacked together in an array. This technique was combined with optical scan tiling, which uses a scanning mirror to combine the signals from the SLMs, thus filling in any gaps in the physical tiling array. Finally, the researchers sped up the full-color video playback using powerful graphics processing units. This combination of technologies produced one high-resolution, full-parallax moving hologram displaying 45 billion pixels per second.

"We increased the resolution of the holographic

display system by 24 times," states Xu. "The full-color 3D holographic video plays at a rate of 60 frames per second, so it appears seamless to the human eye."

Potential applications of the new technique include 3D entertainment and medical imaging. However, new SLM devices with a smaller pixel size, higher resolution and faster frame rate are required before large-scale 3D holographic video displays can become reality.

More information: Xu, X., Liang, X., Pan, Y., Zheng, R. & Lum, Z. A. Spatiotemporal multiplexing and streaming of hologram data for full-color holographic video display. *Optical Review* 21, 220–225 (2014).
dx.doi.org/10.1007/s10043-014-0032-y

Provided by Agency for Science, Technology and Research (A*STAR), Singapore

APA citation: Creating high-resolution full-color moving holograms in 3-D (2015, February 4) retrieved 12 September 2015 from http://phys.org/news/2015-02-high-resolution-full-color-holograms-d.html

This document is subject to copyright. Apart from any fair dealing for the purpose of private study or research, no part may be reproduced without the written permission. The content is provided for information purposes only.

Cave automatic virtual environment

A **cave automatic virtual environment** (better known by the acronym **CAVE**) is an immersive virtual reality environment where projectors are directed to between three and six of the walls of a room-sized cube. The name is also a reference to the allegory of the Cave in Plato's *Republic* in which a philosopher contemplates perception, reality and illusion.

1 General characteristics of the CAVE

A CAVE is typically a video theater sited within a larger room. The walls of a CAVE are typically made up of rear-projection screens, however flat panel displays are becoming more common. The floor can be a downward-projection screen, a bottom projected screen or a flat panel display. The projection systems are very high-resolution due to the near distance viewing which requires very small pixel sizes to retain the illusion of reality. The user wears 3D glasses inside the CAVE to see 3D graphics generated by the CAVE. People using the CAVE can see objects apparently floating in the air, and can walk around them, getting a proper view of what they would look like in reality. This was initially made possible by electromagnetic sensors, but has converted to infrared cameras. The frame of early CAVEs had to be built from non-magnetic materials such as wood to minimize interference with the electromagnetic sensors, obviously the change to infrared tracking has removed that limitation. A CAVE user's movements are tracked by the sensors typically attached to the 3D glasses and the video continually adjusts to retain the viewers perspective. Computers control both this aspect of the CAVE and the audio aspect. There are typically multiple speakers placed at multiple angles in the CAVE, providing 3D sound to complement the 3D video.

2 Technology

A lifelike visual display is created by projectors positioned outside the CAVE and controlled by physical movements from a user inside the CAVE. A motion capture system records the real time position of the user. Stereoscopic LCD shutter glasses convey a 3D image. The computers rapidly generate a pair of images, one for each of the user's eyes, based on the motion capture data. The glasses are synchronized with the projectors so that each eye only sees the correct image. Since the projectors

The CAVE

are positioned outside the cube, mirrors are often used to reduce the distance required from the projectors to the screens. One or more computers drive the projectors. Clusters of desktop PCs are popular to run CAVEs, because they cost less and run faster.

Software and libraries designed specifically for CAVE applications are available. There are several techniques for rendering the scene. There are 3 popular scene graphs in use today: OpenSG, OpenSceneGraph, and OpenGL Performer. OpenSG and OpenSceneGraph are open source; while OpenGL Performer is free, its source code is not included.

CAVELib is the original application programmer's interface (API) developed for the CAVE(TM) system created at the Electronic Visualization Lab at University of Illinois Chicago. The software was commercialized in 1996 and further enhanced by Mechdyne Corporation. The CAVELib is a low level VR software package in that it abstracts for a developer window and viewport creation, viewer-centered perspective calculations, displaying to multiple graphics channels, multi-processing and multi-threading, cluster synchronization and data sharing, and stereoscopic viewing. Developers create all of the graphics for their environment and the CAVELib makes it display properly. The CAVELib API is platform-independent, enabling developers to create high-end virtual reality applications on Windows and Linux operating systems (IRIX, Solaris, and HP-UX are no longer supported). CAVELib-based applications are externally configurable at run-time, making an application executable independent of the display system.

Mechdyne's Conduit is a commercial software package

that makes any existing 3D OpenGL application (like CATIA, Pro/E, Unigraphics...) work directly in a CAVE, without any source code modification. Working like an OpenGL driver, it takes the commands of the existing application, streams them on a PC cluster, and changes the camera so that the viewpoint is dependent on the tracking system.

VR Juggler is a suite of APIs designed to simplify the VR application development process. VR Juggler allows the programmer to write an application that will work with any VR display device, with any VR input devices, without changing any code or having to recompile the application. Juggler is used in over 100 CAVEs worldwide.

CoVE is a suite of APIs designed to enable the creation of reusable VR applications. CoVE provides programmers with an API to develop multi-user, multi-tasking, collaborative, cluster-ready applications with rich 2D interfaces using an immersive window manager and windowing API to provide windows, menus, buttons, and other common widgets within the VR system. CoVE also supports running X11 applications within the VR environment.

Equalizer (software) is an open source rendering framework and resource management system for multipipe applications, ranging from single pipe workstations to VR installations. Equalizer provides an API to write parallel, scalable visualization applications which are configured at run-time by a resource server.

Syzygy (software) is a freely-distributed grid operating system for PC cluster virtual reality, tele-collaboration, and multimedia supercomputing, developed by the Integrated Systems Laboratory at the Beckman Institute of the University of Illinois at Urbana–Champaign. This middleware runs on Mac OS, Linux, Windows, and Irix. C++, OpenGL, and Python applications (as well as other regular computer apps) can run on this and be distributed for VR.

Avango is a framework for building distributed virtual reality applications. It provides a field/fieldcontainer based application layer similar to VRML. Within this layer a scene graph, based on OpenGL Performer, input sensors, and output actuators are implemented as runtime loadable modules (or plugins). A network layer provides automatic replication/distribution of the application graph using a reliable multi-cast system. Applications in Avango are written in Scheme and run in the scripting layer. The scripting layer provides complete access to fieldcontainers and their fields; this way distributed collaborative scenarios as well as render-distributed applications (or even both at the same time) are supported. Avango was originally developed at the VR group at GMD, now Virtual Environments Group at Fraunhofer IAIS and was open-sourced in 2004.

CaveUT is an open source mutator for Unreal Tournament 2004. Developed by PublicVR, CaveUT leverages existing gaming technologies to create a CAVE environment. By using Unreal Tournament's spectator function CaveUT can position virtual viewpoints around the player's "head". Each viewpoint is a separate client that, when projected on a wall, gives the illusion of a 3D environment.

Quest3D A real-time 3D engine and development platform, suitable for CAVE implementations.

Vrui (Virtual Reality User Interface) is a development toolkit that handles real-time rendering, head tracking, etc. in multi-display environments such as the CAVE. 3DVisualizer, LidarViewer, and several other software packages were developed using Vrui to provide visualization tools for specific data types. These tools have been publicly released with continuing development by the Keck Center for Active Visualization in Earth Sciences. Oliver Kreylos maintains Vrui documentation and source code on his website.

inVRs The inVRs framework provides a clearly structured approach for the design of highly interactive and responsive VEs and NVEs. It is developed following open-source principles (LGPL) easy to use with CAVEs and a variety of input devices.

VR4MAX is a package for real-time 3D rendering and development of interactive 3D models and simulators based on Autodesk 3ds Max content. VR4MAX Extreme supports multi-projection for CAVE implementations and provides extensive tracking support.

Cave5D is an adaptation of Vis5D to the CAVE. It enables users to interactively explore animated 3D output from weather models and similar data sets.

EON Icube is a hardware & software package developed by Eon Reality that uses PC-based technology to create a multi-sided immersive environment in which participants may be completely surrounded by virtual imagery and 3D sound. The Icube software supports edge blending and the capability to create full quad buffer stereo images in 3D.

libGlass is a general purpose distributed computing library, but has been used extensively in distributed computer graphic applications. There are many applications running at the five-sided CAVE. For example: astronomic application, arcade-like flight simulator and OpenGL demos.[1]

TechViz XL is a commercial software package that makes any existing 3D OpenGL application (like CATIA, Pro/E, Unigraphics...) work directly in a CAVE, without any source code modification. Working like an OpenGL driver, it takes the commands of the existing application, streams them on a PC cluster, and changes the camera so that the viewpoint is dependent on the tracking system.

P3D VirtualSight is a software solution designed to provide an immersive, photorealistic 3D experience of Digital Aspect Mockups on a 1:1 scale. P3D Virtual Sight supports multiple stereoscopic display modes. It can be interfaced with various tracking systems and can power configurations such as multi-screen devices, image walls based on juxtaposed projections, CAVE systems, or Head Mounted Displays.

Vizard (software) is a multi-purpose virtual reality development platform by WorldViz for building, rendering, and deploying 3D visualization & simulation applications in stereoscopic multi-display environments such as the CAVE. The software lets users control 3D content, CAD workflows, rendering clusters, visual displays, motion tracking, and user interaction from one single platform. A joint solution with SensoMotoric Instruments also allows to incorporate eye tracking.[2]

Quazar3D Immersive (software) commercial software package for building and managing immersive digital environments including CAVE, PowerWalls, cylindrical projection systems, etc. The key feature is a powerful management console for easy configuration of the whole rendering cluster . Features such as VRPN, quadbuffer stereo, hardware and software synchronization, off-axis stereo for planar and cylindrical projections are supported.

Dice by Immersion is an acronym for Digital Immersive and Compact Environment. This is an affordable Premium turnkey CAVE-type solution developed by Immersion SAS (Fr), including hardware (screens, mechanics, projectors, tracking, workstation...), software suite (Middle VR and Unity) and Services (3-year warranty: parts & labour and consumables included).

3D Virtual Spaces by Satavision are CAVE-type solutions including both the hardware and the software developed by Satavision Ltd. The 3D Virtual Spaces are built to customer specific requirements and the content the customer wishes to use is converted into a CAVE compatible stereoscopic content. These spaces are used for multiple purposes: as a tool for planning, research or marketing, in educational settings or as an effective way to increase sales.[3]

VisCube by Visbox affordable high performance CAVE systems that fit within existing spaces, eliminating time-consuming and costly room modifications. VisCube CAVE systems are available as either standalone displays or turn-key VR systems with tracking and software. [4]

3 Calibration

To be able to create an image that will not be distorted or out of place, the displays and sensors must be calibrated. The calibration process depends on the motion capture technology being used. Optical or Inertial-acoustic systems only requires to configure the zero and the axes used by the tracking system. Calibration of electromagnetic sensors (like the ones used in the first cave) is more complex. In this case a person will put on the special glasses needed to see the images in 3D. The projectors then fill the CAVE with many one-inch boxes set one foot apart.

The person then takes an instrument called an "ultrasonic measurement device" which has a cursor in the middle of it, and positions the device so that the cursor is visually in line with the projected box. This process can go on until almost 400 different blocks are measured. Each time the cursor is placed inside a block, a computer program records the location of that block and sends the location to another computer. If the points are calibrated accurately, there should be no distortion in the images that are projected in the CAVE. This also allows the CAVE to correctly identify where the user is located and can precisely track their movements, allowing the projectors to display images based on where the person is inside the CAVE.[5]

4 Applications

The concept of the original CAVE has been reapplied and is currently being used in a variety of fields. Many universities own CAVE systems. CAVEs have many uses. Many engineering companies use CAVEs to enhance product development.[6][7] Prototypes of parts can be created and tested, interfaces can be developed, and factory layouts can be simulated, all before spending any money on physical parts. This gives engineers a better idea of how a part will behave in the product in its entirety. CAVEs are also used more and more in the collaborative planning in construction sector.[8]

The EVL team at UIC released the CAVE2 in October 2012.[9] Similar to the original CAVE, it is a 3D immersive environment but is based on LCD panels rather than projection.

5 See also

- Virtual Reality

6 References

[1] "libGlass - Images". Libglass.sourceforge.net. Retrieved 2014-08-04.

[2] "and WorldViz bring Eye Tracking to Virtual Reality". Smivision.com. 2013-05-28. Retrieved 2014-08-04.

[3] "Satavision". Satavision. Retrieved 2014-08-04.

[4] "CAVE Automatic Virtual Environment". Visbox. Retrieved 2015-09-18.

[5] http://inkido.indiana.edu/a100/handouts/cave_out.html

[6] "Virtual reality in the product development process". Tandfonline.com. 1970-01-01. Retrieved 2014-08-04.

[7] *Product Engineering: Tools and Methods Based on Virtual Reality*. 2007-06-06. Retrieved 2014-08-04.

[8] Nostrad (2014-06-13). "Collaborative Planning with Sweco Cave: State-of-the-art in Design and Design Management". Slideshare.net. Retrieved 2014-08-04.

[9] EVL (2009-05-01). "CAVE2: Next-Generation Virtual-Reality and Visualization Hybrid Environment for Immersive Simulation and Information Analysis". Retrieved 2014-08-07.

7 External links

- Carolina Cruz-Neira, Daniel J. Sandin, Thomas A. DeFanti, Robert V. Kenyon and John C. Hart. "The CAVE: Audio Visual Experience Automatic Virtual Environment", *Communications of the ACM*, vol. 35(6), 1992, pp. 64–72. DOI:10.1145/129888.129892

- Carolina Cruz-Neira, Daniel J. Sandin and Thomas A. DeFanti. "Surround-Screen Projection-based Virtual Reality: The Design and Implementation of the CAVE", *SIGGRAPH'93: Proceedings of the 20th Annual Conference on Computer Graphics and Interactive Techniques*, pp. 135–142, DOI:10.1145/166117.166134

8 Text and image sources, contributors, and licenses

8.1 Text

- **Cave automatic virtual environment** *Source:* https://en.wikipedia.org/wiki/Cave_automatic_virtual_environment?oldid=681663496 *Contributors:* Skysmith, Saltine, Robbot, RedWolf, Alexf, Loremaster, MFNickster, CesarFelipe, Chmod007, CALR, Discospinster, Bender235, Goplat, Stesmo, DrDeke, Polarscribe, Eile, Stephan Leeds, Waldir, Toussaint, Galwhaa, Terryn3, Rjwilmsi, Greerandrew, FlaBot, Imnotminkus, WulfTheSaxon, Dialectric, Joshdboz, Nick, Bnew, Tony1, Mysid, PTSE, Bayerischermann, Dfa~enwiki, JLaTondre, Garion96, Veinor, Davepape, DXBari, Ariedartin, Erhardt, Twinbee~enwiki, Colonies Chris, Bdebackere, Whibbard, Cybercobra, Black Butterfly, DMacks, Anjow, UncleDouggie, Epbr123, Rduff-moliver, Magioladitis, Pennel, Sarahj2107, Balloonguy, Pramual, Drke11ogg, Russell Freeman, Tvashtri, Jeff G., Rand1732, Pperron, Dany beaumon, Andqso, HoloVis, Sunriseshell, VanWoods, Regnirpsj, JuLun~enwiki, HampKaushik, Ceeller, Addbot, Zishan6772, Robert The Rebuilder, Yobot, Ogil, AnomieBOT, Zangai~enwiki, Piano non troppo, Cwfraresso, Control.valve, Ktrippen, J04n, GrouchoBot, Damian Ward, Myitguy, Dysumner, 3dcreator, Tim1357, Lotje, Reaper Eternal, DARTH SIDIOUS 2, Dnthong, Hschilling, Wikipelli, ZéroBot, VRsim, Pachetra, Bomazi, ClueBot NG, Charlesineden, BG19bot, 10eonreality, Neøn, Vikleviki, Robocon1, BattyBot, Googlymoogly64, Wakawakwiki, Mesmer8, Schlueer, 2009sdk2009, Joysword417, Scott Hrs, Ma11dejra, Ineractive3D, Cavefin, Aeromedman, IamM1rv and Anonymous: 199

8.2 Images

- **File:CAVE_Crayoland.jpg** *Source:* https://upload.wikimedia.org/wikipedia/commons/6/6d/CAVE_Crayoland.jpg *License:* Public domain *Contributors:* own work (self-photograph using timer) *Original artist:* User:Davepape
- **File:Commons-logo.svg** *Source:* https://upload.wikimedia.org/wikipedia/en/4/4a/Commons-logo.svg *License:* ? *Contributors:* ? *Original artist:* ?
- **File:Question_book-new.svg** *Source:* https://upload.wikimedia.org/wikipedia/en/9/99/Question_book-new.svg *License:* Cc-by-sa-3.0 *Contributors:*
Created from scratch in Adobe Illustrator. Based on Image:Question book.png created by User:Equazcion *Original artist:* Tkgd2007

8.3 Content license

- Creative Commons Attribution-Share Alike 3.0

www.ingramcontent.com/pod-product-compliance
Lightning Source LLC
Chambersburg PA
CBHW080907170526
45158CB00008B/2025